A KNOWLEDGE OF LIVING THINGS

WITH THE

LAWS OF THEIR EXISTENCE.

A

KNOWLEDGE OF LIVING THINGS,

WITH THE

LAWS OF THEIR EXISTENCE.

BY

A. N. BELL, A.M., M.D.,

LATE P. A. SURGEON, U. S. NAVY; PHYSICIAN TO BROOKLYN CITY HOSPITAL.

New York:

BAILLIERE BROTHERS, 440 BROADWAY.

LONDON:—H. BAILLIERE, 219 REGENT STREET.

PARIS:—J. B. BAILLIERE ET FILS. MADRID:—C. BAILLY-BAILLIERE.

1860.

THOMAS HOLMAN, PRINTER,
Corner of Centre and White Streets, N. Y.

PREFACE.

THIS book is offered to amateur readers and teachers of Natural Science under its proper title.

I coincide with the great majority of Physicians, and many Teachers who have tried it, that it is impracticable to so abridge Human Physiology as to render it entertaining to the general reader or comprehensible to the academic student.

Physiology comprehends the history of all living things, and *Human* Physiology is the highest point of that Science, because it is the history of the most complex of all living beings.

All Sciences have their rudiments, and as well might we undertake to acquire a knowledge of the higher branches of mathematics without first studying arithmetic, or reading without knowing the letters, as Human Physiology without that knowledge which constitutes the rudiments of life.

If we would have a simple Physiology, it must be the Physiology of a simple organism—an organism which consists of but few parts, and whose life is maintained by simple processes; and if we would extend this knowledge, we must do it in the same manner as we pursue other sciences—proceed step by step from the lower to the higher branches. Such knowledge as this constitutes the RUDIMENTS OF PHYSIOLOGY.

The circumstances which have led to this book, were in the first place dependent upon a fondness for physiological studies; secondly, an abundant opportunity in various climates to promote those studies; and thirdly, a desire to inculcate that knowledge of living things which is of utility in all departments of human industry. My first effort to inculcate this knowledge was in a course of lectures on Physiology and Hygiene, first delivered a little more than a year ago at the Packer Collegiate Institute of this city. The immediate inducement to compile a book from those lectures, has been brought about by the desire of some of the eminent teachers who heard them.

In submitting my manuscript to the Publishers, they wisely decided that to print it with *fac-simile* drawings from those used at my lectures, might make it a work for connoisseurs, but the expense of the plates would destroy its prospect for extensive reading and for academic instruction. They therefore suggested that it would be best to replace some of the drawings by figures selected from the finely illustrated works on their shelves. Selections were made from the latest London Editions of Carpenter's *Comparative Physiology;* T. Rymer Jones' *Natural History of Animals;*

The Ocean, by P. H. Gosse; and *The Plant*, by Schleiden. These selections required corresponding quotations or deductions from the context instead of my own matter.

The works of OWEN, CARPENTER, FARADAY, PARIS, MUDIE, DANIEL, J. K. MITCHELL, DUNGLISON, SCHLEIDEN, LIEBIG, T. RYMER JONES, and GOSSE, have been my principle authorities for reference.

A. N. BELL.

BROOKLYN, March, 1860.

CONTENTS.

CHAPTER XI.

CHAPTER XII.

CHAPTER XIII.

CHAPTER XIV.

CHAPTER XV.

CHAPTER XVI.

CHAPTER XVII.

CHAPTER XVIII.

CHAPTER XIX.

CHAPTER XX.

CHAPTER XXI.

CHAPTER XXII.

CHAPTER I.

THE KINGDOM OF NATURE.

1. When we direct our attention to the infinity of objects in the universe, we observe that there are general resemblances among all the works of Nature. This infinity of objects is only a repetition of a certain number of general characters endowed with special attributes.

2. Among animals,—the horse, the dog, the sheep, etc., are easily distinguished; among vegetables,—the oak, the pine, the peach, and the apple, are equally easy of distinction; and among minerals,—the metals, lime, granite, slate, etc., all have their characteristics. By studying a specimen of each of these varieties in Nature, the powers of the memory acquire some knowledge of the whole, and this knowledge enables us the more easily to obtain that information which is of most practical utility, namely, to avoid what is hurtful, and to use that which is most beneficial to us. Such knowledge as this begins with the first exercise of reason, and the pursuit of it is so gratifying to the reasoning faculties, that the labor of ages has now nearly completed an arrangement of the materials of the universe, under two grand divisions into which the Kingdom of Nature has been divided.

3. The Ancients divided the domain of Nature into *three* classes:—Minerals, Vegetables, and Animals. But modern discovery has demonstrated so close a connection between Vegetables and Animals as to justify

their consideration under one head, and, owing to their possessing *organs*, or instruments adapted to the performance of particular *functions*, they may be classed together as Organic Bodies. While, on the other hand, those bodies which have no organs, are, by common consent, called Inorganic Bodies. These two classes may be distinguished from each other by numerous peculiarities.

4. Of all the characteristics which distinguish the two great divisions of natural bodies, the most absolute and the most palpable, is that which is drawn from the manner of *growth* and of *nourishment.*

5. Inorganic bodies grow only by accretion, that is to say, by the accession of new layers to their surface ; whilst organic bodies, in virtue of their vital powers, receive and assimilate the substances which contribute to their growth, by penetration. In animals and plants, nutrition is the effect of an internal mechanism ; their growth is developed from within. In minerals, on the contrary, growth cannot claim the name of development ; it goes on externally, by successive additions of new layers ; it is the *same being*, assuming other dimensions, whilst the organic body in its growth is *renewed.*

6. Another remarkable difference between organic and inorganic bodies, consists in the homogeneousness of the latter, and the heterogeneous or compound nature of the former. Let a block of marble be broken: each piece will be perfectly similar to the rest; there will be no differences among them, except such as relate to size and shape. Break down the fragments and each grain will contain particles of carbonate of lime, which will be throughout the same. On the other hand, the division of a vegetable or an animal, shows parts heterogeneous or dissimilar. In different parts there will be

found muscles, bones, arteries, veins, nerves ; blossoms, leaves, bark, pith, etc. And of composition, organic bodies cannot live or exist in their natural condition, unless solids and liquids enter at once into their composition. The coexistence of these is necessary, and living bodies always contain a liquid mass more or less considerable, which is incessantly agitated by the motion of the solid and living parts. It is, indeed, impossible for life to exist, without an intricate combination of solids and fluids. The water which exists in combination with mineral substances is not a necessary part of them—they can exist without it. The water of crystallization though intimately combined, and rendered solid in crystallized substances, is not necessary to the existence of the substance, though perhaps essential to the particular form exhibited. And these inorganic and homogeneous substances, when resolved by decomposition into their last elements, are formed by particles similar to one another, and possess a great simplicity of internal arrangement.

7. The *size* and *shape* of inorganic bodies, depend upon the number of particles which can be brought together, and the part to which they are applied. They may be great or small, regular or irregular.

8. On the other hand, organic bodies attain a certain size and shape, by a process of growth, and in all cases there is an appropriation of the elements to particular parts, maintaining a due proportion between the stem and the root, the limb and the trunk. Each vegetable and animal has its own size and shape by which it may be known.

9. There is also a difference in the *degree* of composition. From the mineral to the vegetable and from the latter to the animal body, there are gradations of com-

plexity. The multiplicity of the elements entering into the composition of the animal body, accounting in a measure for its increased tendency to alteration of parts.

10. Inorganic compounds are often binary, as the greater part of saline substances. Sometimes they are ternary but seldom quaternary ; while the most simple organic body rarely contains less than four elementary principles.

11. The *active properties* of inorganic and organic bodies are also quite different. There are certain forces which affect all matter, but in addition to these are certain other forces peculiar to organic bodies, and therefore called *vital forces*, these are *contractility* and *sensibility*. All animals and many plants, maintain certain motions under the influence of stimuli, which can under no circumstances be excited in inorganic matter.

Fig. 1.

Dionæa Muscipula.—Venus' Fly-Trap.

12. In the Venus' Fly-trap, for instance, the contact of any substance with one of the prickles which stand upon each lobe of the leaf, will occasion the closure of the lobes together, by a change taking place in the leaf-stalk. And *contractility* to a less extent is a property of various other plants and all animals.

13. *Sensibility*, is a term applied to powers exercised through the instrumentality of a nervous system, and is alone applicable to animals.

14. In *origin* and *end* there is also a manifest difference. Organic bodies spring from a germ which is at first a part of another being, and organic bodies only can die. They all have a duration determined by their nature, and this duration is very different from that of inorganic bodies, which is proportioned to bulk and density.

15. Finally, these differences may be comprehended under a great variety of circumstances, as in the following

TABLE

Of the differences between Inorganic and Organic Bodies.

1.—Composition.

Inorganic Bodies.	*Organic Bodies.*
Homogeneous, atoms are independent of each other.	Heterogeneous, atoms are dependent upon each other.
Simplicity and constancy of chemical properties.	Complexity and variation of chemical properties.
Never present a union of gaseous, liquid, and solid parts.	Always present a union of fluid and solid parts.
Capable of being decomposed and recomposed.	Capable of decomposition, but incapable of recomposition.

16. 2.—Form.

Of Inorganic Bodies.	*Of Organic Bodies.*
Variable.	Constant.
Usually angular.	Usually rounded.

17. 3.—Active Properties.

Common to both.—Gravitation, Repulsion, and Chemical affinity.
Peculiar to Organic Bodies.—Contractility and Sensibility.

18. While men, in the progress of Science, were examining the forms and other qualities of the bodies

around them, they could not avoid noticing the motions and changes going on among them. And it was thus that the grand discovery was made that there were resemblances in all the works of Nature.

19. Museums of Natural History have been formed in modern times, which contain specimens of almost every object that may be included in the two great classes in Nature, so that now within the limits of a school cabinet and garden, we are able to survey the whole material of the universe.

20. The result of the countless observations and experiments made through a series of ages, shows that all the changes, motions, and phenomena of natural objects are merely a repetition and mixture of a few simple forms or kinds of motion and change, which are as constant and regular in every case, as are the motions of the heavenly bodies. The expressions which describe these changes, are denominated LAWS OF NATURE.

21. Acquaintance with the Laws of Nature has been very slow in its progress, owing to the complexity of ordinary phenomena, which is produced by several laws operating together, and under various circumstances. But enough has been learned, for us to perceive that the great universe is as simple and harmonious as it is immense; and that the Creator, instead of interposing separately or miraculously—in the common sense of the word—to produce every distinct phenomenon, has willed that all should proceed according to a few general truths. And there is nothing in Nature so truly miraculous and adorable as that the endless and beneficent variety of results which we witness, should spring from such simple means.

22. In times of ignorance, men naturally regarded every occurrence which they did not understand, that

is to say, which they could not refer to a general law, as arising from a direct interference of Supreme power; and thus for many ages, and among some nations still, the wind and weather, eclipses and earthquakes, and many diseases—particularly those of the mind, were or are still accounted miraculous. Hence arose among heathen nations many ceremonies, and even barbarous sacrifices, for propitiating or appeasing their offended deities; but founded on expectations no more reasonable than if we should now pray to have the day or the year made shorter; or to have a coming mortality averted, while we indulge the conditions which are certain to lead to it.

23. The beauty and regularity of the appearances in Nature have in all ages attracted the attention of observers. The ancient Greeks embodied their impressions in the word by which they designated the visible world, which they called COSMOS, to denote at once its order and beauty. While the Latins styled the world MUNDUS, to express their sense of its surpassing loveliness.

24. From the very beginnings of philosophy the sources of natural phenomena have been the subject of contemplation and research. Ancient philosophers referred them to various transformations of certain elements, and these they attempted to classify as Air, Water, Earth, and Fire, which they deemed undecomposable and indivisible, and therefore the first general elements of the universe. But Science in its progress has been continually widening the dominion of law, and what the ancients deemed to be simple and undecomposable, has long since been resolved into simpler substances, again and again simplified, until Chemistry now estimates no less than *sixty-five* simple substances which

are regarded as indivisible and undecomposable. And there are combinations of these sixty-five elements and of the properties possessed by them, so numerous that they can never be numbered, and so diverse that they can never be classified; yet there is certain space in Nature's domain allowed for them all, and the play of their several qualities, each one being endowed with peculiar properties in reference to all the others. And it is only by the harmonious action of these properties that the universe is sustained.

25. The gaseous, fluid, and solid forms of matter with heat, are analogous to the primary division by the ancients into Air, Water, Earth and Fire, and these, collectively considered, embrace all the varieties of form of which matter is susceptible.

26. The qualities of physical objects all proceed from the same source, and have certain operations in relation to organic beings, yet all in harmony; so that we have a complexity of result from the simplest means, which is eminently characteristic of the works of God.

27. "The world," says Farady, "with its ponderous constituents, dead and living, is made up of natural elements endowed with nicely balanced affections, attractions, or forces. Elements the most diverse, of tendencies the most opposed, of powers the most varied; some so inert that, to a casual observer, they would seem to count for nothing in the grand resultant of forces; some, on the other hand, endowed with qualities so violent, that they would seem to threaten the stability of creation; yet when scrutinized more narrowly, and examined with relation to the parts they are destined to fulfill, are found to be in accordance with one great scheme of harmonious adaptation. The powers of not one element could be modified without destroying at

once the balance of harmonies, and involving in one ruin the economy of the world."

28. The unity of purpose with which things work in Nature, that are to our observation, when we think of them singly, most opposite, is one of the most delightful rewards of observing them in their combination.

29. In every part of creation, latent powers are at work giving intimations of their existence by signs which cannot be mistaken, yet not deigning to afford any insight into their nature, further than that they are subservient to the law of relative adaptation, and to demonstrate that of such materials as have been revealed to man and others subject to like conditions, the Creator hath made the universe.

30. The first general conception with which we should begin the study of vital phenomena, is the mutual dependence of all things, which, in their coördinate whole compose the system of Nature,—a dependence which requires for each being, the simultaneous existence of all the others. Thus a vegetable derives its nourishment from inorganic bodies, and these become so changed by the influence of vegetable life, as to be fit food for animals.

31. While the doctrine of vital phenomena, or Physiology, is an independent science, resting upon truths of its own, which it draws from the observation of those actions, which, in their aggregate connection constitute life, it is enriched by facts furnished to it by collateral sciences. And the admirable relations existing between the conformation of our parts, and the external objects to which they are applied, carries along with the study of Physiology a lively interest in all that pertains to it.

32. By the aid of Natural Philosophy, Chemistry, and

Mathematics, the physiological relations of the human system may be calculated with such precision, and laid down with such accuracy, that the functions of our organs come to be regarded as models of the most ingenious productions of art.

33. How the atoms of matter approach each other, and cling together to form masses, which are solid, liquid, or aëriform, can only be learned by a knowledge of *Physics* or *Natural Philosophy.* And if we would obtain a correct knowledge of the nature of the solids, fluids, and gases, and of the manner in which organized substances are constantly passing from the one to the other of these different forms of matter, we must study *Chemistry.* To *express* these different physical relations and affinities of Natural Philosophy and Chemistry, *Mathematics* become also an essential study of the Physiologist.

34. But of all collateral studies, *Anatomy* is that which will enable us to obtain the best appreciation of the beauties of living phenomena. Anatomy is, indeed, a part and parcel of Physiology, bearing the same relation to it as geography does to history. A knowledge of the situation, the size, the form, the relations, and the structure of organs being indispensable for a perfect understanding of their functions. It is necessary, therefore, to combine so much of the study of these sciences as may be necessary to render our own theme intelligible

CHAPTER II.

OF THE PROPERTIES OF MATTER.

35. Certain substances, when placed in contact, exhibit a proneness to combine with each other, or to undergo a decomposition, while others may be mixed most intimately without change. This attraction, by which the particles of dissimilar matter are united into a compound, is called *chemical affinity*. And a chemical compound thus formed, partakes of the properties of neither of its constituent elements, but acquires a new and distinct character.

36. The particles of a body in the solid or liquid state, also exhibit an attraction for each other, which is the force of *cohesion*. And again, different kinds of matter have an attraction for each other, which is probably of the same nature as cohesion, but because the bodies are different in their nature, this attraction is called *adhesion*. These differences are illustrated by the action of glue, wax, putty, and other cements, in connecting bodies together. In detaching glue, or putty, from the surface of glass, the latter is sometimes broken or torn off by the glue or putty, the *adhesion* being stronger than the *cohesion* of the glass. But both of these properties materially differ from chemical affinity—*they effect no change in the properties of bodies.* They combine different kinds of matter together, but they do not alter their nature.

37. Chemical affinity may be distinguished from all other properties of matter, by the deep and complete

change of characters which follow its action; briefly, *it is the force by which new substances are generated.*

38. In the generation of compound substances by chemical affinity, the elements which enter into them, always unite in fixed and invariable proportions. For instance, water is composed of eight parts of oxygen to one of hydrogen; and in no other proportions could these elements form water. This condition of matter has led to the adoption of what is called the *Atomic theory.* This theory is, that the ultimate particles, called *atoms* (from *atomos*, indivisible), possess, in different substances, different weights; and that when bodies unite, it is the atoms which combine with one another.

39. When substances unite in two or more proportions, the several proportions with which one body combines with another, are the ratio of 1, 2, 3, 4, 5. And the number which represents the proportion with which one body is capable of combining with another, is also the representative of the proportion in which the same body combines with every other; that is to say, that bodies combine in reciprocating ratios. For instance, eight parts of oxygen unite with one of hydrogen, with six of carbon, and with fourteen of nitrogen; and the six of carbon is the exact quantity for union, with one of hydrogen, and eight of oxygen, and fourteen of nitrogen; thus showing that the weights of the combining proportions of these bodies are equivalent one to another. But these numbers do not represent the absolute weights, they only show the *relative* weights or combining proportions. Sometimes, however, these relative weights represent correct estimates of the real weights, as oxygen and hydrogen, the combining proportion of oxygen being eight times

heavier than hydrogen, though hydrogen occupies twice the volume. This example also shows that the equivalent bulks of bodies bear no relation to their equivalent weights.

40. Bodies also combine by volumes, in ratio, similar to what they do in weights. That is to say, substances, when aëriform, unite in volumes which are equal; or when unequal, the larger volume multiplies regularly, is double, triple, or quadruple the smaller, admitting of no fractional parts.

41. Elementary substances arrange themselves into new compounds, according to the degrees of their reciprocal attraction or chemical affinity, and they are incapable of farther changes under the same conditions.

42. But there is still further provision for the formation of compounds, by a law of chemistry called *Catalysis*, from *kata* downwards, and *luo*, I unloose. Catalysis *is a property of matter by which, when two or more substances are in contact, any motion occurring among the particles of the one may be communicated to the particles of the other.* In virtue of this property, in many cases, bodies which have but little tendency to unite with each other, when brought in contact with a substance for which neither has affinity, enter into combination. But it is natural to suppose, that while the elements of every such compound remain the same, their affinities for each other might be just as completely satisfied by a different arrangement of the atoms. Such compounds may, therefore, be changed to others, by a change in the direction of the atoms, of which they are constituted.

43. The most important means of exciting catalysis in bodies, and enabling them to rearrange themselves under new forms, consists in bringing them in contact with a substance undergoing decomposition. If,

for instance, a little sugar be put in contact with yeast, it unites with the elements of an equivalent of water, and divides itself into *two* equivalents of alcohol, and *four* equivalents of carbonic acid. To comprehend this, it is necessary to understand certain formulæ which chemists use instead of writing out the equivalent proportions of compound bodies. Sugar consists of *twelve* equivalents of carbon, *eleven* of hydrogen, and *eleven* of oxygen. To express this in formula, it may be written $C_{12} H_{11} O_{11}$. Alcohol is composed of *four* equivalents of carbon, *six* of hydrogen, and *two* of oxygen, and *two* equivalents of alcohol may be expressed by 2 ($C_4 H_6 O_2$). Carbonic acid is formed by an equivalent of carbon, and two equivalents of oxygen, so that to express *four* equivalents of carbonic acid, write 4 ($C O_2$). The yeast, possessing the power of separating the compact into which carbon, hydrogen, and oxygen had entered to form sugar, these several elements now being under the control of new influences, take a totally different course, divide into two cliques, and the result is alcohol and carbonic acid, substances as totally different from sugar as if they were constituted of different elements. Almost the converse of this takes place by the same law, in the processes of germination and malting. Starch ($C_{29} H_{20} O_{20}$) being converted into sugar by means of *diastase*, a substance which accompanies starch in the grain.

44. Starch is universally diffused through the vegetable kingdom, and its great purpose is, the nutrition of the growing body before that is capable of obtaining food for itself. But before starch can be applied to the nutrition of a germinating seed, it must be changed into sugar. And this change is effected by the *presence* of diastase, which seems to have no other purpose.

The slow decomposition of diastase, excites in a great amount of the atoms of starch the necessary degree of motion for their rearrangement, and the presence of water (H O) furnishes the necessary elements for the formation of sugar. The nature of diastase, other than what is here given, is not known ; it contains nitrogen, and it is probably in virtue of this element, that the change of starch into sugar is effected, but diastase, isolated and pure, has never been obtained.

45. Although the vital principle of organic bodies exercises great power over chemical forces, it only influences these forces by directing the way in which they operate, and not by changing the laws to which they are subject. Inorganic matter entering into organic bodies continues subordinate to the same laws which regulate independent chemical action, the chemical forces exercised under the influence of life producing the same results as if no such influence existed.

46. The passage of matter from a state of motion to that of rest, or the process of formation and growth, in vegetables and animals, goes on in the same way; in both the same cause determines the increase of the mass. And this constitutes true vegetative life which is carried on without consciousness.

CHAPTER III.

OF INORGANIC BODIES.

47. In order to keep up the phenomena of life certain matters are required, but how the supply is continued and how the material is disposed of are important questions. Wonders surround us on every side. The formation of a crystal, is no less incomprehensible than the formation of a leaf or of a muscular fibre, and it is the extent of our understanding to know upon what conditions these formations depend. *How* they are formed we shall never know.

48. *Oxygen*, *Carbon*, *Hydrogen*, and *Nitrogen*, are essential elements of all organic bodies, and it is only by a knowledge of the properties of these elements that we can contemplate the phenomena of life.

49. OXYGEN is a colorless gas, destitute of odor and taste. It is the most generally diffused element in nature; for, besides constituting one-fifth of the bulk of the atmosphere and one-third of the bulk of water, it is a component of almost all the earths and minerals found on the surface of the globe, and is a universal constituent of all animal and vegetable substances. Oxygen is superior to all other elements in the extensive range of its affinities, and it is the agent employed in effecting the union and disunion of a vast number of compounds.

50. The first necessities of animal life, are nutritious matters and oxygen. And at every moment of existence the animal is taking oxygen into its system, by means of

the organs of respiration, and no pause is observable while life continues.—The quantity of oxygen respired by an adult man, in one year, amounts to about 750 lbs., or nearly 33 ounces per day; yet we find his weight at the beginning and end of the year nearly or quite the same.

51. What, it may be asked, has become of the enormous weight of oxygen introduced? No part of it remains in the system. It has been returned to the atmosphere in the forms of carbonic acid gas, and the vapor of water.

52. Since no part of the oxygen taken into the system is again given off in any other form but that of a compound of carbon or hydrogen, and as the carbon and hydrogen are supplied in the food, it is clear that the amount of nourishment required by the animal body must be in direct ratio to the quantity of oxygen respired.

53. As all elementary bodies combine in equivalent proportions, the amount of carbon required to convert 33 ounces of oxygen into carbonic acid gas, may be easily estimated. Carbonic acid consists of one equivalent of carbon and two equivalents of oxygen, or 6 parts of carbon produce exactly 22 parts of carbonic acid. Hence the daily amount of carbon required to convert 33 ounces of oxygen into carbonic acid is 9 ounces.

54. CARBON. This substance is of universal extent, and exists under a great variety of circumstances. In its purest state, it is represented in the diamond. Anthracite coal, plumbago, lampblack, wood-charcoal, coak, ivory-black, etc., are familiar examples of nearly pure varieties.

55. Carbon surpasses all other substances in its affinity for oxygen, with which it unites in two propor-

tions forming the gaseous compounds known as carbonic acid and carbonic oxide. It is evident that a cause must exist which prevents the increase of carbonic acid in the atmosphere, by removing that which is constantly forming, and also of replacing oxygen which is removed by the processes of organization and putrefaction. Both of these causes are united in the functions of vegetable life.

56. Carbon and the elements of water form the principal constituents of vegetables ; the quantity of other substances existing in very small proportion. The formation of the principal component substances of vegetables, is attended with the separation of the carbon of the carbonic acid from the oxygen ; this carbon enters into combination with the elements of water, and the oxygen is returned to the atmosphere. For every equivalent of carbonic acid thus decomposed, two equivalents of oxygen are liberated, carbonic acid consisting of one equivalent of carbon and two equivalents of oxygen. This remarkable property of plants is easily demonstrated.

57. The leaves and other green parts of a plant absorb carbonic acid, and emit oxygen, and they possess this property quite independently of the plant; for if, after being separated from the stem, they are placed in water containing carbonic acid, and exposed in that condition to the sun's light, the carbonic acid is, after a time, found to have entirely disappeared from the water. And if the experiment be conducted under a glass receiver filled with water, the oxygen emitted from the plant may be collected and examined. When no more oxygen is evolved, it is a sign that all the carbonic acid is decomposed; and the operation will commence anew, if carbonic acid be again added to the water.

58. It is thus that in the economies of Nature, plants not only afford the means of nutrition for the growth and the continuance of animal organization, but they likewise furnish that which is essential for the support of the important vital process of respiration; for besides separating all noxious matters from the atmosphere, they are an inexhaustible source of pure oxygen, which supplies the loss that the air is constantly sustaining. Animals, on the other hand, expire carbon, which plants inspire; and thus the composition of the medium necessary for the existence of both, namely, the atmosphere, is maintained in a state of purity.

59. It is as impossible for the organic functions of plants to cease for a single moment without death, as it is for the respiratory functions of animals. During the night, or when plants are in the shade, carbonic acid accumulates in all parts of their structure, and if their growth is vigorous, carbonic acid is sometimes emitted; but from the instant that the rays of the sun fall upon them, the exhalation of oxygen commences.

60. Plants improve the air by the removal of carbonic acid, and by the renewal of the oxygen. And thus it is that vegetable culture heightens the healthy state of a country, and that a previously healthy country may be rendered sickly by the discontinuance of cultivation.

61. Hydrogen. This is an insipid, inodorous, and colorless gas, which will not support life nor combustion, though it is preëminently inflamable and a principal constituent of ordinary flame. It is the lightest of all substances, being sixteen times lighter than oxygen, and fourteen times lighter than the atmosphere.

62. When mixed with a large quantity of air, hydrogen may be taken into the lungs without inconvenience and it is in no way deleterious, but as it does not sup-

port respiration, an animal placed in it soon dies of suffocation. It is, however, a very important constituent of organic matter. It has a strong affinity for oxygen with which it unites and forms water, and in this form it is a constituent of all living bodies; but when life ceases it acquires a still stronger affinity for nitrogen. The whole phenomena of decay depend upon the exercise of these affinities, and many of the processes engaged in the deciduous function of plants originate in the partial exercise of affinities provided for death and decay.

63. As already stated, the solid parts of plants chiefly consist of carbon and the elements of water. Under the influence of solar light, healthy plants decompose water, by assimilating the hydrogen with carbonic acid; and by this means oxygen is evolved. It has been estimated that from this source, each acre of land which produces 10 cwts. of carbon, gives annually to the atmosphere 865 lbs. of free oxygen gas.

64. NITROGEN. This gas is distinguished by a comparative want of properties. Its principal characteristic is an indifference to all other substances, and an apparent reluctance to enter into combination with them. When forced to do so, it seems to remain in combination by the power of inertness only, and very slight forces will effect the disunion of its feeble affinities. It is frequently called *azote*, from two Greek words which signify inability to support life. A burning taper is instantly extinguished in this gas, and an animal soon dies in it, not because the gas is injurious, but from the privation of oxygen. Yet it forms four-fifths of the bulk of the atmosphere, and enters into the composition of all organic bodies. It is not, however, always found in vegetable substances, though it is well known that no plant can attain maturity without the presence of matter containing nitrogen.

65. Of animals, no part which possesses motion is destitute of nitrogen. The chief ingredients of human blood contain nearly 17 per cent. of nitrogen, and it is an essential element of food for all the purposes of nutrition. Yet it seems to take no other part in the functions of life than mere presence, the vital processes requiring this, however, for their healthy exercise.

66. When the mysterious principle of life has ceased to exercise influence, nitrogen assumes a peculiar activity, and becomes a promoter of death and decay, by escaping from the elements which have held it in abeyance. Its utility now becomes manifest. It is emphatically the element of death. During life, it is subject to the control of the vital powers. But no sooner does life cease, than nitrogen acquires a strong affinity for hydrogen, and combining with it a new compound—*ammonia*—is formed ; oxygen is thereby disengaged, and a new set of affinities begins, converting what would otherwise be a state of rest into one of commotion. Fermentation is excited, and decomposition proceeds uninterruptedly to a complete transformation.

67. *Ammonia* is the last product of decay and putrefaction. It is a gaseous *compound*, and consists of one equivalent of nitrogen, and three of hydrogen. It is a colorless gas, of strong and pungent odor, familiar in spirits of hartshorn, and is distinguished as the volatile alkali. Ammonia is exceedingly soluble in water, and cannot therefore remain long in the atmosphere, as every shower of rain condenses it, and conveys it to the surface of the earth. Hence rain-water always contains more or less ammonia, and it is this which gives to rain-water the apparent sensation of *softness* generally experienced. The ammonia which is removed from the atmosphere by rain, watery vapor, etc., when

in contact with other substances, enters into new affinities, by means of which the nitrogen is returned to the atmosphere.

68. All the innumerable products of vitality now assume their original form. And thus death—the complete dissolution of an existing generation—again becomes the source of life for a new one.

CHAPTER IV.

AIR.

69. In the changes that take place in nature, the state of air is of the first importance. It is the first change of matter essential to the constitution of organic bodies. Air may be defined to be matter so overcome by tendency to motion, that it has no cohesion, and none of the common properties of matter, except gravitation. Yet air may mean anything, or all things,—for the elements of which all things are composed, may exist in the state of air, and this state is the beginning of everything new, and the end of everything old. But the type and ordinary example of air, is the atmosphere.

70. Air is the most delicate and the most powerful of all substances. While it yields to the touch of the sunbeam, it cleaves rocks asunder, and demolishes whole countries in earthquakes. It is nicer in the detection of pressure than any instrument ever devised, and no instrument can compare with an air-thermometer in measuring the degree of temperature. The air is fine beyond all sensation, yet it is the immediate object of all the senses. It is the air which the eye sees, the ear hears, the nose scents, and the finger touches. Without the air, we know nothing of sight, hearing, smelling, touch, or the sensations of space. The eye would be immediately destroyed, were it deprived of air ; its fluids would burst their sacks, and there would be an end of its curious structure and all of its power.

71. In relation to the ear, air is the only instrument of sound, and if it were not present, all nature would be dumb. To the touch: could one clasp in the hand any substance, without intervening air, never again could the fingers be unclenched. And the sole of the foot, if once planted on the surface of the earth without air between, would remain as fixed as the rooted sapling. If air did not intervene between substances said to be in contact, there would be no need of fixation points. There would be no change, no life.

72. Air is both the pathway and the carrier of organic nature. It is no less yielding to everything else, than sensitive to its own integral motions. If anything advances, the air moves off before it to make room; and if anything recedes, the air follows to support it. If anything is heated above its wont, the air ascends with the excess of heat; and if anything is cooled overmuch, the air closes in upon it to impart warmth and protection. No matter how great or how small the object, how long or how short the distance, the air is sensible to the very smallest degree of action, and equally capable of adaptation to the most powerful. It surrounds all things, regulates all things.

73. The sphere of matter which surrounds the liquid and solid surface of the earth, is called *atmosphere*, from two Greek words, which signify *sphere of vapor*. It chiefly consists of two ingredients, *oxygen* and *nitrogen*, in the proportion, by volume of twenty-one per cent. of the former to seventy-nine per cent. of the latter, and these proportions have been found to exist whencesoever taken, whether from the summits of the highest mountains, or from the sandy plains of Africa.

74. Besides oxygen and nitrogen, the atmosphere always contains a variable quantity of watery vapor and *carbonic acid gas*. The proportion of this latter,

varies from 6.2 as a maximum, to 3.7 as a minimum in 10,000 volumes. Its proportion near the surface of the earth is greater in summer than in winter, and during night than during day. It is also rather more abundant in elevated situations, as on the summits of high mountains, than in the plains; which distribution of this gas proves, as will be seen by and by, that the action of vegetation at the surface of the earth, is sufficient to limit the proportion of carbonic acid in the atmosphere.

75. *Ammonia* is also constantly present in the air, though in an exceedingly small proportion. Such are the constituents of the atmosphere necessary to plants and animals. And although other gases and vapors enter the atmosphere, they are speedily decomposed and dispersed, and cease to exist.

76. The lower surface of the atmosphere is equal to about 200,000,000 of square miles, and if it were everywhere of the same density as at the surface its entire height would not exceed five miles; but the atmosphere decreases in density in a geometric progression for equal heights, and it is estimated by meteorologists to expand to the height of about fifty miles. On account of its elasticity, its volume is always inversely proportioned to the pressure upon it: its elasticity directly as the pressure.

77. The pressure of the atmosphere at the level of the sea, is equal to about fifteen pounds on every square inch of surface, so that the body of a man of ordinary stature—whose superficial surface may be estimated to be fifteen square feet—sustains a pressure of 32,400 pounds! This pressure is not felt, in consequence of the cavities of the body and of the bones being also filled with air and fluids, equally elastic with the exter-

nal air; and these counterbalance the outward pressure, so that no inconvenience is experienced. To illustrate this, if an animal be placed under the receiver of an air-pump, and the air be exhausted, the air within the body being no longer counterbalanced by the pressure of that without, expands ; the animal appears inflated and soon dies.

78. The weight of the atmosphere is capable of sustaining a column of water thirty-four feet high; or one of mercury thirty inches high, as in the barometer.

79. The range of the barometer varies from 28 to 31 inches ; and if the changes in the pressure of the atmosphere varying between these extremes be not extremely sudden, the human system is not liable to suffer from them ; but if we rapidly descend to great depths below the surface of the earth, or ascend to a great height in the air, our functions become deranged by the disturbed equilibrium of pressure, and disease is likely to ensue.

80. The indisposition experienced under such circumstances has been often experienced by baloonists and others. A German traveler (Meyen) describes the effects produced on a party of persons who hurriedly ascended the mountains of Peru :—"We were tormented with a burning thirst, which no drink was able to assuage : a slice of watermellon which we had brought with us was the only thing we could relish, whilst our people ate garlic and drank spirits, maintaining that this was the best way to guard against the effects of the journey. We kept on ascending till two o'clock in the afternoon. We were already near the summit of the mountain, when our strength at once abandoned us, and we were overtaken by the disease *Sorocco.* The nervous feverishness under

which we had suffered from the first, had been gradually becoming worse and worse, and our breathing became more and more oppressed; fainting, sickness, giddiness, and bleeding at the nose came on; and in this condition we lay a considerable time, until the symptoms grew milder from repose, and we were able to descend slowly."

81. The author is familiar with a sad case which occurred under his own observation, during the Mexican war, 1848. Lieut. R., of the U. S. Navy, formed one of a party, who hurriedly ascended the mountains of Mexico, starting from Vera Cruz. Though well when he set out, and with no known predisposition to pulmonary disease, he was at the end of three weeks time, after having gained a high altitude, seized with severe hemorrhage from the lungs, from the effects of which he never recovered.

82. It is the pressure of the atmosphere at its ordinary level, that prevents the escape of the fluids contained in the bloodvessels; and if this pressure be largely diminished, hemorrhages are apt to occur from those parts of the body where the vessels are least protected, as in the windpipe and its branches into the lungs.

83. It must be apparent, therefore, that sudden transitions from a dense to a rarer atmosphere, is exceedingly unfavorable to such as are predisposed to lung affections; and that for all, such transitions should be made with great moderation, so as to enable the system to adapt itself to them.

84. Heat. When Air is rapidly and forcibly compressed, heat becomes manifest. The pressure, density, and *temperature* of the air, are all by the nature of its constitution so nicely balanced, that the least change in any one of these conditions is instantaneously fol-

lowed by a corresponding change in the others; and its freedom of motion enables it to make an immediate adjustment of its parts. There are few or no causes of disturbance arising from pressure in the air, because the only pressure which it has is its own weight, or pressure downwards toward the earth. And the pressure of foreign substances mixed with the air are always very circumscribed, and exercise but little influence. But the pressure of the aerial particles is, nevertheless, of such importance as to be the chief cause of all the motions which take place in the atmosphere.

85. The elevation of temperature which results from the friction of hard bodies, is analogous to the forcible compression of air; such bodies, like air in a comparative state of rest, producing heat in proportion to the degree of motion excited. And as Air was the first condition of all hard bodies, it is easy to perceive that heat is simultaneous in both origin and existence with air.

86. We have seen that air pervades all space. Through this undulations are propagated, which give impressions of heat. A hot solid body possesses the property of exciting undulations of heat in the air, which spread on all sides around it like the waves from a pebble thrown into still water. Sounds are propagated in the same manner; and all the experiments on reflection and concentration of heat, may be imitated by means of sound.

87. It is thus seen that when the effect of heat on air is made so manifest as to be readily perceptible, that it is in all respects analogous to motion, the most prominent characteristic of air. The air is in a constant state of motion, and this *motion causes heat.*

88. Heat and motion are convertible one into the other. The powerful mechanical effects produced by the elasticity of aeriform bodies affords abundant illustration of this ; and the production of heat by friction, shows with equal clearness that motion may be converted into heat. It is clear, therefore, that heat is a property of the air.

89. FIRE, or combustion, is heat concentrated, supported by the oxygen of the atmosphere. When motion or heat is very much intensified—as in active friction —fire is the result.

90. ANIMAL HEAT. Heat in organic bodies is the necessary result of transformations or changes which are being effected for other purposes. And it is produced most abundantly in the higher animals, in those which are considered warm blooded, because in them the functions of organic life are most active.

91. Heat is produced, however, to a certain extent by all animals, and occasionally by plants. But when heat is produced by plants, they exist under the same conditions as animals, that is to say, when the compounds of which they were constituted are being restored to the condition of inorganic matter.

92. LIGHT. Light is supposed to consist of undulations excited in the ethereal medium pervading all space, and filling up the intervals between the particles of ponderable bodies. Moreover, the particles of this ether are supposed to vibrate, not in the direction of the ray, like particles of air in conveying sound, but in planes at right angles to the length of the ray, like the transverse vibrations of a stretched cord.

93. The processes by which light is maintained, are not clearly understood ; there can be little doubt, however, that light is dependent upon the action of heat,

in which the carbon and hydrogen of living bodies are being returned to the atmosphere as carbonic acid and water. This conclusion is sustained by analogy, in that, intense heat or combustion always produces light.

94. ELECTRICITY. Like light, electricity owes its origin to heat or motion. Heat will generate electricity when applied to two dissimilar metals in contact, or, as in the usual manner of generating it, by the friction of two dissimilar substances. But the usual and powerful source of electricity, is chemical action or motion; there being probably no instance of chemical change or decomposition, without giving rise to electricity. Light, too, by means of the chemical change which it produces, may generate electricity.

95. When the change of a body from a solid to a liquid form, or from a liquid to a gaseous is attended with chemical decomposition, the motion or heat thereby generated always produces electrical disturbance. The ordinary process of growth in organic bodies is constantly producing such changes as these, under circumstances peculiarly favorable to the excitement of electricity. There can, therefore, be no difficulty in accounting for the production of electricity in the processes of organization.

96. Thus Heat, Light, and Electricity are *properties* of the AIR, and necessary to its perfect operation. And in this sense, air is an element in nature, not chemically inseparable, but as a *whole*,—and an element essential to the functions of organization. Air is so constituted that it holds all matter captive; but the captivity is only that the purposes of matter may thereby be accomplished. For the moment that the proper ingredients of any compound come in contact in the right proportion and under the requisite circumstances, the

hold is relinquished ; the ingredients instantly act, and the new compound is formed with as much ease and certainty as if it had existed from the beginning.

97. Heat is as powerful in escaping, and allowing matter which it holds in the state of air to act in the formation of a new substance, as it is in the suspension of properties which have already served their purpose and occupy materials to no use. The solvent power of heat, which loosens the firm cohesion of the diamond with as much ease and certainty as it melts ice into water, or the sunbeams into those tints of color that enliven the face of nature, overcomes all, but destroys nothing.

98. All the changes in the air are accompanied by changes in its properties, of which heat is predominant. And all subsequent elements are consequences of aerial changes and condensations, even as air is the first condensation of ether, the apparent nothing of infinite matter.

CHAPTER V.

WATER.

99. Air being the first degree of ethereal condensation in the order of necessary elements for organic beings, WATER may be regarded as the second. It, like air, consists of invisible atoms, but they adhere more strongly to each other, are more condensed than are the atoms of the atmosphere. In it oxygen obtains a preponderance ; water consisting by weight of eight parts of *oxygen* to one part of *hydrogen*.

100. The nature of water is, that in large or small quantities, it always tends to reproduce the globe, or to form drops. It has a gaseous effort towards the formation of a general globe, and also an effort to a central or individual globe. It possesses, therefore, an effort unto form, while it is constantly relapsing into formlessness. This oscillation between form and formlessness is the conception of fluidity, which is essentially different from that of gaseity. Water seeks to produce fluidity everywhere, by the formation of globules, or *solution*,—its chief function.

101. Every solid formation has come out of water, and without water there would be no solid, no organic being, no life. By solution, solids are reduced to the primary condition, and are then capable of reassuming new forms. The process of solution, is the process of becoming water, not by agglutination, but by liberation of fixity. No process of solution is conceivable

without oxygen, and during every solution, the principles of water enter into a state of tension with each other. If the solution be very heterogeneous, the oxygen and hydrogen separate, and the water is decomposed.

102. The heterogeneous nature of compound bodies renders water incessantly subject to decomposition. In the laboratory of the clouds, in the pulsating vessels of plants and animals, water is constantly being decomposed, and its principles reappropriated to the never-ending changes of the universe.

103. As we do not see the particles of the atmosphere, as a whole, the particles of its two chief ingredients, oxygen and nitrogen; nor the particles of water, oxygen and hydrogen, which the air takes up in the process of evaporation; we cannot know the nature of the agency by which any of these are held together. The cohesion of the particles in the entire substance, as air, is indeed not only small, but absolutely negative, and entirely obedient to the property of heat.

104. When water passes into a state of vapor, or becomes endowed with that dispersive motion of its particles which sends it invisibly through the atmosphere, it is changed to a state very much resembling air; and thus it ascends in consequence of heat—the dispersive property of the air. But although the vapor is invisible, we are not to suppose that it necessarily enters into chemical combination with the air, in such a manner as that the two form one substance. On the contrary, it is only dispersed *through* the air mechanically, and rises by the law of gravitation, because the quantity of it which is contained in any given bulk is less by weight than the quantity of air contained in the same space.

105. Water, when in the most minute state of division, and ascending in the air in little invisible drops, and so thin that the slightest cobweb would sink in it like a stone, is, notwithstanding, as perfectly water as when it rolls in the flood of a river, or spreads in the ocean ; and it is just as subservient to all the laws of its nature in the one situation as in the other.

106. The atmosphere is usually and justly styled the breath of every living thing. But it is something more than that. Were the atmosphere to suspend its general evaporative power for ever so brief a period of time, the very beginnings of life would be cut off, and its fountains dried up. All the waters which the rivers of the world roll to the sea, all that slumbers in ponds, and expands in lakes, all that is caught in fountains, drawn from wells, or in anyway appropriated to the processes of the arts, or the purposes of life ; all that supplies drink to the whole animal race, and is breath, and life, and clothing, and habitation to innumerable tribes ; all that waters the fields, and sustains the existence of every vegetable, from the moss on the rock to the monarch of the forest ; all that enters into the structure of plants and animals, or which bears their more immediate nourishment on its tide, or cleanses, softens, and comforts them by its ablution ; nay, all that enters into those stones and gems which glitter so much, is brought on the wings of the wind, mounts up through the viewless air, and the more vigorously that the countless thousands of active powers, natural or artificial, are working, the more abundantly does the air supply them with natures most abundant—next to air itself, most refreshing and valuable production.

CHAPTER VI.

THE EARTH.

107. The Earth is a magazine of inorganic matters, which are prepared by vegetation to suit the purposes destined for them in the nutrition of animals. In composition these matters consist of a preponderance of carbon, together with a large proportion of oxygen and a small proportion of hydrogen and nitrogen. The atoms of these are so arranged that unlike Air or Water, the gravity of ether is fixed, and there results the highest degree of ethereal condensation or immobility of atoms, with a tendency to a single direction, the centre; this arrangement constitutes *cohesion of particles,* the distinctive characteristic of *solid bodies.*

108. There was a period in the Earth's existence when no vegetation clothed its solid crust, in which no animal lived, when no *humus* could possibly be present. "The Earth was without form and void, and darkness was upon the face of the deep. The Spirit of God moved upon the waters." The dense cloud was swept away, the mountains rose, and the Earth rolled in Light! Then came into action the decomposing influence of the atmosphere. Penetrating to the very bottom of the deep crevices still hot with the volcanic throws of internal fire; the Air, expanding in the power of its might, hurled from their lofty summits great blocks of stone, to be dashed into pieces in their descent. In process of time, vegetation gradually developed in such vast quantity, in such gigantic

luxuriance, that the same, being buried layer after layer by subsequent revulsions, might be prepared and preserved for the present economy of the world.

109. The gigantic ferns, palms, reeds, and other similar plants, which covered the Earth's surface in primeval periods, belong to a class of plants which have the power to dispense with nourishment from the soil. The constitution of these first plants was similar to those which are now produced from bulbs and tubers, which are chiefly nourished by absorbing moisture and carbonic acid from the atmosphere. Such plants, while young, live upon substances contained in their seeds, and after their exterior organs of nutrition are formed, they require little or no food from the earth. Plants of this character are at present ranked amongst those which do not exhaust the soil. They are distinguished by the inconsiderable development of their roots and the enormous expansion of their leaves.

110. By the decay of these first plants, the soil in which they grew first become supplied with *humus*, the product of vegetable decay, which was necessary for the development of succeeding generations of plants.

111. During the whole life of a plant, transformations of the elements which constitute it are incessantly taking place. In consequence of these transformations, gaseous matters are given off by the leaves and flowers, solid matters are deposited in the bark, and fluid substances containing carbon are excreted by the roots and absorbed by the soil. By the expulsion of these matters which have served their purpose in the processes of nutrition, the earth receives again with usury the carbon which it at first yielded to young plants as food, in the form of carbonic acid. Plants

thus give back more carbon to the earth than they receive from it.

112. Decay of vegetable matter is no less important in another point of view; for by means of its influence, the oxygen which plants retained during life, is again restored to the atmosphere. But the decomposition of vegetable matters is effected in various ways, and followed by variable results. Woody fibre kept under water, or in dry air, may last for thousands of years without suffering any appreciable change; but when it is exposed to the air, in the moist state, the oxygen of the air combines with it to form humus, and the mass is gradually changed into *coal.*

113. Coal formations are exclusively of vegetable orgin, formed apparently from the destruction of plants, in the revulsions of the earth. The enormous quantities of timber drifted by some of the great rivers of the world into the present ocean, render it probable that coal formations may now be going on in some parts of the sea. It may also be inferred that coal beds were first formed in the neighborhood and often upon the verge of extensive tracts of dry land. Others appear to have been elevated and singularly contorted, as if thrown up by volcanic forces.

114. The coal mines of most ancient formation may be distinguished from those of later date, by the absence of *roots* in the vegetable remains which constitute them. The tubers and bulbs which pertain to plants of the lowest degree of organization, are exceedingly perishable, and the first parts to suffer decomposition. But in plants more highly organized, such as oaks and other trees, which are the product of later ages in the earth's history, the roots are tough and fibrous, and no more perishable than the stems and leaves. There-

fore, in coal mines of more recent formation, the roots of the trees which formed them are always present.

115. But all the carbon which enters into the formation of coal mines, must have been originally in the atmosphere as carbonic acid, in which form it was assimilated by the plants which constitute these formations. It follows from this, that the atmosphere must be richer in oxygen than it was in the early periods of the earth's history. And the increase, is in exact proportion to the quantity of carbon and hydrogen contained in the coal formations.

116. During the formation of every thousand cubic feet of coal, the atmosphere must have received nearly two thousand cubic feet of oxygen, produced from the carbonic acid assimilated, and also over two hundred cubic feet of oxygen resulting from the decomposition of water. In former ages, therefore, the atmosphere must have contained less oxygen and a much larger proportion of carbonic acid, than it does at the present time, a circumstance which accounts for the greater richness and luxuriance of the earlier vegetation.

117. But a certain period must have arrived in which the quantity of carbonic acid contained in the atmosphere, experienced neither increase nor diminution in any appreciable quantity. For if it received an additional quantity to its usual proportion, an increased vegetation would be the natural consequence, and the excess would then be speedily removed. And, on the other hand, if there was less than the normal quantity of carbonic acid, the process of vegetation would be retarded, and the proportion would soon attain its proper standard.

118. The quantity of coal accumulated in the superficies of the earth's surface exceeds computation. But

it has been estimated that the annual consumption in Europe alone exceeds 34,000,000 of tons. And geology shows that even if the consumption of coal should increase, the store will certainly last five hundred years longer. Such a store of coal corresponds to about 13,000,000,000 of tons of carbon, abstracted from the air of that region, before the earth was fitted for the habitation of Man !

119. Plants and animals require substances of different kinds ; some are formed solely of carbon and the elements of water ; others contain nitrogen. But besides these elements which are essential to all degrees of organization, certain other substances are requisite for the formation of organs destined for special functions. Indeed it is probable that the whole category of elementary substances are adaptable to some forms of organization.

120. The arrangement of all the different parts and portions of the Earth is the consequence of the same Intelligent Power that adapts all natural agents to the support of life under the most pleasurable conditions. If we pass over a beautiful and varied country, the scenery may fill our minds with the most pleasing emotions ; but when we are informed that it is all owing to the most violent convulsions which have upheaved the crust of the Earth, and that that is a proof of wise and benevolent design, we are either inspired with wonder or incredulity. But the geologist estimates the importance of the cause that leads to such results. He knows that if the strata which compose the exterior of the Earth had never been thus upheaved, no beds nor channels would have been formed for the superficial waters except those uncertain excavations consequent on its own feeble motion, and that

under such circumstances the Earth never could have been adapted to its occupants. And if the eruptions had been less violent, the coals and metals which are now so near the surface would have remained beyond the utmost researches of man.

121. Such conclusive facts cannot fail to evince an intelligent design, and the choice of the best means; and they are instances of the provision for the welfare of what was to follow in producing such an arrangement of causes, as is best calculated to secure a state of things most suited to sustain the permanence and happiness of animate existence.

122. The mineral elements most commonly met with in organic bodies, are—Phosphorus, Lime, Sulphur, Iron, Magnesia, Chloride of Sodium or common Salt, Manganese, Silica, and Alumina. Some of these appear to be introduced merely to answer certain mechanical or chemical purposes, but others intimately combine and arrange themselves with Oxygen, Carbon, Hydrogen and Nitrogen, into the formation of all the parts essential to a perfect organization.

123. No food can serve for the nourishment of an organic body unless it contains all the elements of which the body is composed. Vitality is the power which each organ possesses of constantly producing itself; but for this purpose the organ requires a supply of material which contains the elements of its substance, and these elements must be capable of undergoing transformation. All the organs together cannot *generate* a single element—carbon, nitrogen, or a metallic oxyd—they must all have existed before.

124. The inorganic earthy salts and metallic oxyds which are found in the remains of vegetable decomposition are the foundation of all the changes that take place

in organic bodies, and these are the powers we reach in our ultimate researches in the processes of organic bodies. It is, therefore, incorrect to consider mineral bodies as inert. All things are active ; that is to say, they exercise certain properties peculiar to themselves.

125. The ashes of plants and the bones of animals are nearly identical in composition, and consist of the mineral substances above named, together with the essential elements to their combination, and these earthy matters are indispensable for the development of organic bodies.

CHAPTER VII.

LIFE.

126. The elements of the world, in the order of creation, remain not here in a state of useless suspense, but, by the special endowment of the Creator, they become living beings.

127. The phenomena of development, growth, sensation, decay and death, and many others, belong to LIFE ; but as Life only occurs in material structures which subsist in obedience to the laws of physics and chemistry, it is truly a superstructure on these laws, and cannot be studied independently of them. Indeed, the greater part of the phenomena of organic beings are chemical and physical phenomena, modified only by an additional principle, *Life.*

128. LIFE *is the visible expression of that arrangement in nature by which elementary substances are in such wise combined as to form organic bodies.* Health, disease, and death, are the functions of the ratio of this arrangement.

129. Life is at first composed of a small number of phenomena, but under its influence organs and instruments are developed and multiplied, and the properties which characterize life and bear witness of its presence, being at first obscure, become more and more manifest, increasing in number as in energy and power. And, as from the lowest beings we ascend to Man, the field of Life's existence is continually enlarging. The organs

or instruments are many times multiplied, and the whole system becomes more and more wonderful in its adaptations to manifold purposes. Yet every living being is perfect in itself; each one is constituted most favorably to the purpose it is designed to fulfill in the Kingdom of Nature. And from the lowest vegetation to the very crown and summit of organic bodies, all is equally admirable.

130. But it is only by studying the first conditions of life, that we can appreciate the genaral plan to which each organic being conforms.

131. And as the human race in its progress must pass through definite stages of civilization, so also in the gradual development of organic nature, all depends upon definite physical laws.

132. Man's life is inseparably linked with the plants and animals which coexist with him, and these are the issue of long anticipations and preparations. Not only the comfort, health, and degree of civilization, but the very existence of Man depends upon the state of the earth, the atmosphere of the earth, the climate of the earth, and the productions of the earth. Man is placed in a system where all the changes produced in other objects occur according to a relation existing among the substances changed, and his own organic constitution participates in all these things that surround him. To understand these conditions of our existence, it is necessary to begin at the very germ of organization, and pursue the changes that take place in the nearest approximation to the inorganic material of the universe.

133. The beginnings of all things are small, and the neglect of small things is the grand error in studying life as well as in starting life; and the nearer to the beginning that the deviation from a right course is, the

greater the wrong in the end. Nature never acts on visible matter, she only acts on the aerial, or invisible particles—air with air. Whenever a solid disappears, it passes through a liquid form into a state of vapor or air. And as in the first elements, there are various degrees of condensation and tenacity; the same variations pervade all the works of nature.

134. In magnitude, living beings are subject to the greatest possible variations. Quantities are perpetually divisible, insomuch that if it were possible to do as much with our hands as with the understanding, we should be able to take from any given magnitude a part which should be less than any other magnitude given. There is, indeed, no impossible smallness of bodies.

135. The smallest creatures cognizable to the human senses, by the aid of the microscope, the Infusoria Monads, when powerfully magnified, only appear as mere moving points (Fig. 2, *a*, *d*, *e*). Other species, however, the Monas Termo, *b*; the Uvella, *c*; the Polytoma, *f*, *g*; and the Microgena, *h*, *i*, appear large enough to enable us to obtain some idea of their structure. Some of them are observed to have little appendages resembling the tail of the Tadpole. Distinguished microscopists, who have carefully examined this appendage, inform us that so far from being a tail it is a *proboscis* or mouth, and

Fig. 2.

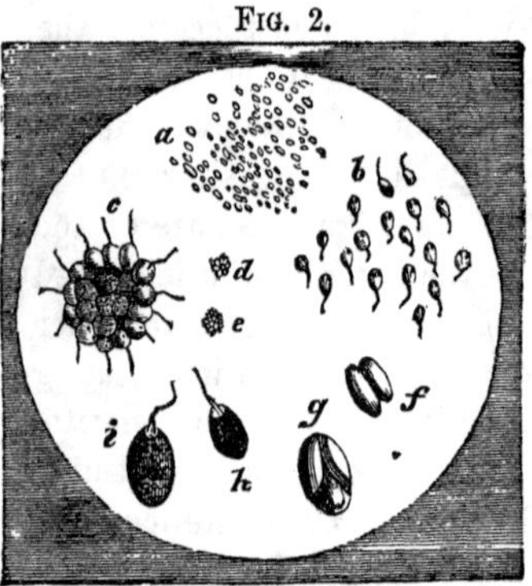

INFUSORIA MONADS.

that it answers the double purpose of capturing food and of scudding the little creature through the water in search of it.

136. Living things which are so small as not to be cognizable to the senses without the aid of a powerful microscope, have their offspring. And these young ones have their little veins and other vessels, and their eyes, so small that no microscope can make them visible. So that we cannot suppose any magnitude so little but that Nature produces a less. Yet there are microscopes so made that the things we see appear more than two thousand times larger than they would appear if we looked upon them with the unaided eye. And it is reasonable to believe that the power of the microscope may be so augmented, that every one of these two thousandth parts will seem more than two thousand times larger! Nor is the smallness of certain bodies to be admired more than the greatness of others, for it belongs to the same Infinite Power as well to augment infinitely, as infinitely to diminish. The majesty of the Creator appears equally in things small or great, and as it exceeds man's comprehension in the immense greatness of the universe, so also does it in the smallness of all the parts.

137. When, from the depths of the sea, a volcano, bursting through the boiling flood, upheaves a scoriaceous rock, producing a new island; or where the united labor of the Lithophytes have piled up their cellular dwelling, after toiling thousands of years, till their coral edifice reaches the level of the sea,—no sooner does the newly-made surface from these causes greet the air, than it begins to be spotted over with the formative processes of life. (Plate 1.) The surface first becomes of a dusty softness, and looks like velvet.

This velvet-like covering gradually assumes a yellowish color, and is divided into various sections, bordered by rings, some single and some double, of different shades, traversed by furrows dividing them into compartments. As time runs on, the spots which first formed become of a darker hue ; from a bright glittering yellow, they turn brown, and gradually to a bluish gray, and then to a dusky black. Meanwhile, new ones are just beginning ; the circles and squares increase in size, and merge into each other, until their lines lose distinctness. But new ones here begin, of dazzling whiteness. And thus one tissue rises above another, in strata of different colors and variable thickness, like embossed work. In this way, as the surface of the rock becomes drier by exposure to air and sunshine, *soil* is made and prepared for vegetative life. This soil speedily becomes covered by a rich green plush, gorgeously arrayed in Nature's favorite green. (Plate 11.)

138. The rich green plush here described, is the first result of the joint administration of the simplest natural elements, and it consists of the lowest condition of living bodies.

139. Life is at first comprised in a little *cell*, or bag, or an aggregation of these, each one of which is capable of separate and independent existence, and capable of multiplying its species, simply by subdivision. The reproduction of species, however, even in the simple degree of vital existence, is usually accomplished by the union of the contents of two cells, and if they are permitted to remain in contact, embedded as they are in a gela-

FIG. 3.

LOWEST TYPE OF LIFE.

tinous substance, forming a single homogeneous mass, they go on multiplying, until they are forced into a linear arrangement, *A*.

Fig. 4.

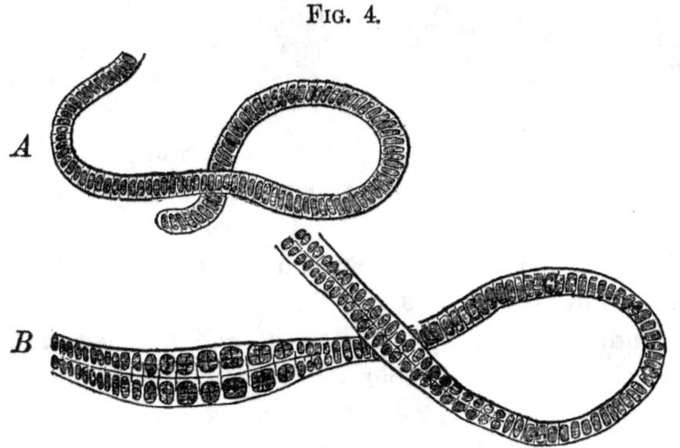

Multiplication of Species in the cell,—The Thallus.

140. The linear series subdivides, first transversely, and afterwards longitudinally, so as at last to produce a thin expansion, called a *Thallus*, which is the first approach towards the formation of a *leaf*, *B*. In the simplest forms of the thallus, there is no difference perceptible in the cells which constitute it, every one of them appearing to have an independent existence ; and, moreover, able to multiply itself, not merely by division, but by the emission of a portion of its contents, inclosed in a cell-wall, or new *germ-cell*, called a *spore*.

141. The spore is analogous to the seed of a more highly organized plant. It is furnished with a fine fringe, or hair, by means of which it is endowed with the power of motion, and so becomes a crawling spore, or what is termed a *Zoospore*.

142. When the Zoospore has separated itself from the primitive cell, it begins to develop itself into a new organization similar to that from which it sprang, by subdivision; but in doing so, it passes through the primitive form.

Fig. 5.

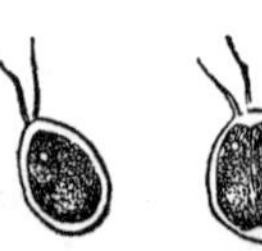
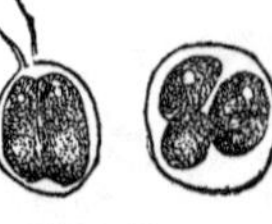
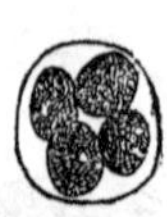

The Zoospore.

143. The perfect organization of any one of the higher plants or animals, is not more dissimilar in form and condition to that of the lowest and simplest member of either the vegetable or animal kingdom, than it is to the germ of its own kind in its first period of existence. And, indeed, when we go back to the beginning of organic structures, or rather, as far back as we can, for even when the powers of the microscope are exhausted, we are still in the midst of the process, we discover that the *simple cell* is that state *in which every living being, both animal and plant, commences life.* And it is not until this cell has grown, or proceeded to a considerable extent, that it can be discovered with certainty, whether it is that of a plant or an animal; the evolution of the first organic germs, proceeding in a perfectly similar manner.

A Fig. 6. *B*

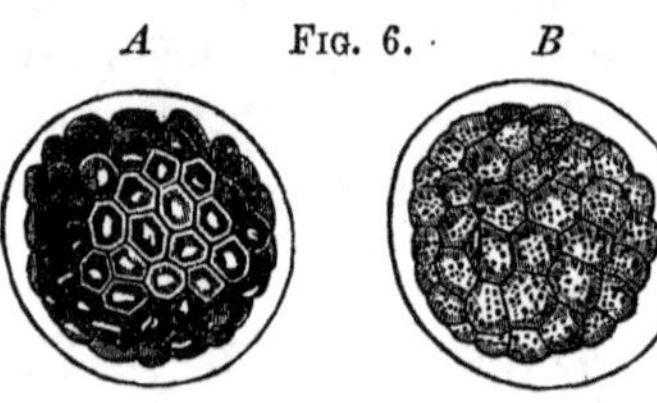

Identity of the Germ Cell.
A, *Animal ovum.* B, *Vegetable Monad.*

144. We have here represented the Mammal ovum, formed by the repetition of the duplicative subdivision

of the germ-cell, which bears a most striking resemblance to the vegetable Monad of the Volvox. But notwithstanding the similarity in the first forms of living beings, every cell, like every individual plant or animal, is the product of a previous organization of the same kind. There being no reason to believe that any living being, no matter how simple, can originate in the combination of inorganic elements, without the intermediation of a substance already endowed with vitality. This substance, however, may not be a definite germ particle, for it oftentimes appears to be nothing else but a fragmentary portion of the liquid contents of some preëxisting cell.

145. But although we may not be able to discern any difference in the *germs* of different orders of living beings, yet there is such a marked diversity in their operation under similar circumstances, that we cannot do otherwise than attribute to them distinct properties. In virtue of these properties each germ develops itself into a structure of its own specific type, whenever it is supplied with the requisite material, and the required forces are brought to bear upon it.

146. The first stage of development consists in the production of the necessary *pabulum* or material out of the inorganic elements which constitute Air, Water and Earth. In this operation the germ particle seems to act in a manner analagous to a *ferment* or leaven, excepting only that the tendency of the change is in the opposite direction from that which takes place under the influence of a ferment. Ordinary fermentation is the decomposition of organic matter in the downward direction, towards inorganic bodies; while the metamorphosis of a *cell pabulum* is in the ascending direction, from inorganic matter up to organic compounds.

147. It is under the influence of warmth and light only that the change of inorganic matter into cell pabulum can take place ; and the amount of the new compound thus generated, as measured by the quantity of oxygen given off in the act of their production, is in exact proportion to the degree in which the decomposition of carbonic acid is promoted by the agency of Light. A common appearance of the organizing process of the simplest forms of living vegetable matter, is that of a green scum, which floats upon the surface of stagnant water on exposure to the sun. This scum disengages bubbles of gas, which, upon analysis, is found to consist of oxygen.

148. Besides oxygen in every such situation where vegetable nutrition is being formed, starch-gum or *dextrine* and *albumen* are present ; the albumen being abundant in proportion to the rapidity of the organizing process, and chiefly administering to the internal contents of the cell, or that part which is most actively concerned in the functions of life. Other products are only met with in a more advanced stage of cell-life, after the cells have ceased to perform active functions, and when their composition is less immediately related to that of the cell-walls. The *dextrine*, or starch-gum, which is always present in much greater abundance than the albumen, is the pabulum out of which the outer wall of the cell is generated, its thickness being augmented from time to time by additional exudations. This distinction of parts, and the appropriation of the *albumen* and *dextrine*, is beautifully illustrated in a section of the leaf of an *Agave* which has been treated with dilute nitric acid.

FIG. 7.

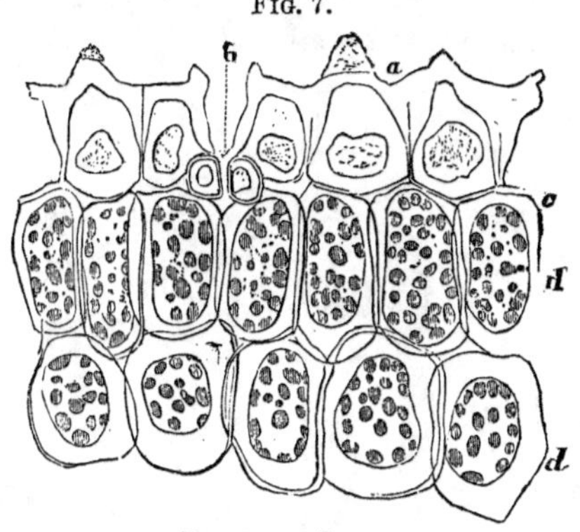

DISSECTED CELLS.

a, epidermic cells or outer covering ; *b*, boundary cells of the *stoma* (holes through the epidermis or outer covering) ; *c*, cells of *parenchyma*, those cells which are condensed into the hard part of the plants ; the inner line at *d*, encloses the *primordial utricles* or most elementary parts ; these constitute the immediate investment of the contents of the cell, and seem to form the starting-point of vital operations.

149. Before the dextrine and albumen are appropriated to the formation of the cell, they are incorporated into a peculiar viscid fluid which has received the designation of *protoplasma*, which appears to be in immediate relation with the production of the living and growing cell.

150. The protoplasma, in addition to the starchy and albuminous compounds, also contains sugar and oily matter. In young cells, the protoplasma fills nearly the entire cavity, certain small spaces only being perceptible, which are occupied by a more watery fluid called *cell-sap*. But with the advance of life, the cell-sap is taken up by the protoplasma and appropriated to the lining of the cell-walls.

151. In this protoplasma, a most remarkable movement of cell-life is often observed, which is known

under the name of *rotation.* It was first noticed in a simple form of the water-plants, the *Nitella flexilis.*

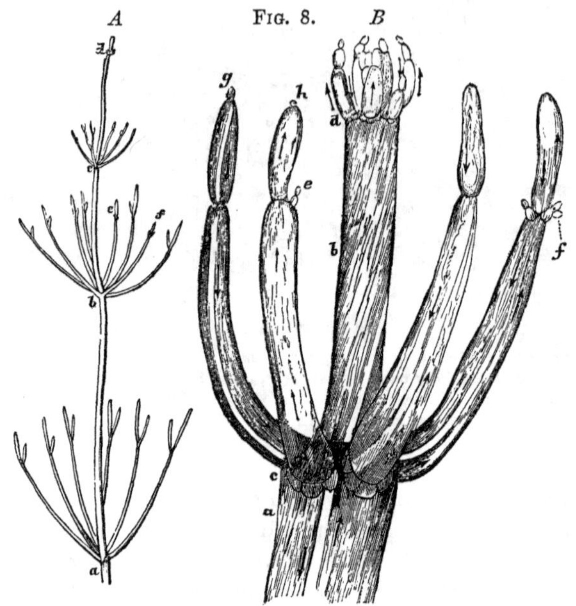

ROTATION.

A, stem and branches of the natural size ; *a*, *b*, *c*, *d*, four verticils of branches issuing from the stem ; *e*, *f*, subdivision of the branches ; *B*, magnified portion ; *a*, *b*, joints of stem ; *c*, *d*, verticils ; *e*, *f*, new cells sprouting from the sides of the branches ; *g*, *h*, new cells sprouting at the extremities.

152. Each cell of the Nitella in the healthy state is lined by a layer of green granules, which covers every part of its walls, excepting two longitudinal lines, that remain nearly colorless (B). Over this green layer a continuous stream of semi-fluid matter is constantly flowing ; the current, passing up one side, changing its direction at the extremity and flowing down the other

side,—the ascending and descending spaces being bounded by transparent lines, as shown in the figure.

153. These currents are directed or controlled by the layer of granules, as may be shown by interrupting them near the boundary between the ascending and descending currents, when a portion of the fluids of the two currents will pass the boundary and intermingle ; but after a time, when the injury has become repaired by the development of new granules, the circulation resumes its regularity—no part of either current passing the boundary.

154. In young cells, rotation may be seen before the formation of the granular lining, and the progress of it may be affected by anything which influences the vital activity of the plant. The moving globules which chiefly consist of starchy matter, are of various sizes and figures, being sometimes very small and definite, whilst at other times they are large and irregular, appearing to be made up of an aggregation of small particles.

155. The currents observed in rotation, are connected with the general activity of the life of the cell, and when the currents stop, the formative powers cease. From the manner in which the currents are connected with the nucleus of the cell, they are evidently the centre of vitality in the plant. And from what has now been shown of the production of vegetable cells, it is easy to perceive that this current is the concentration of the forces diffused through the protoplasma for the production of new cells with like functions.

156. In the further progress of the development of living beings, the order of classification is the next obstacle. This, however, is evolved in its turn. And the Family, Genus, Species, Sex, and Individual in regular succession.

CHAPTER VIII.

DIVERSITIES OF DEVELOPMENT.

PROTOPHYTES.

157. In assigning a place to any living being we have to proceed from the general to the special attributes of every such being. The lower we descend, whether in the animal or vegetable series, the nearer approach do we make to that homogeneousness which is the typical attribute of inorgânic matter, where every particle has all the marks of individuality, and there is no distinction of either tissue or organs. On the other hand, as we ascend the scale of organization, the fabric becomes more and more heterogeneous, the body is divided into organic systems, and these systems again divided into sections.

158. The lowest grades of vegetable life are denominated PROTOPHYTES, from *proto* (first), and *phyton* (plant), or first plants. Their generation is accomplished by a union of the component cells, all of which are alike; but their formation may be accomplished in various ways.

159. The cell of the Protophyte may develop itself as an individual, by appropriating its pabulum to the extension of its walls; or solidify its thin coats by depositing layers on the inside, so as to gradually fill up its cavity; or by filling its cavity with products of various kinds which it secretes from its primitive pabu-

lum. Or, again, it may give origin by its own subdivision to two or more cells, which in their turn may undergo a like multiplication; the life of the first one being thus perpetuated and distributed through a vast accumulation of beings similar to itself.

160. The most common method of multiplication is the subdivision of the original cell into halves. This phenomena occurs in the *Red Snow*. On the freshly fallen snow of the Polar Regions, and on the summits of high mountains, this unicelled plant often tinges the whole snow field of a rosy-red color. This color is produced by a multitude of single cells, and each one of these being capable of self-existence, may be considered a distinct individual.

Fig. 9.

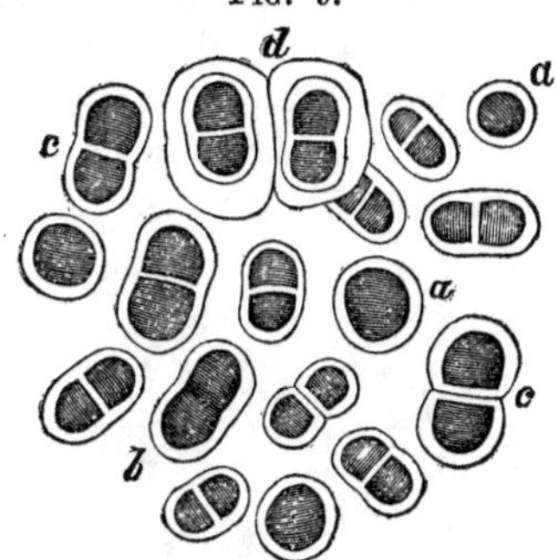

Hœmatococcus Binalis, or Red Snow, in various stages of development.

a, a, simple round cells; *b*, elongated cells, the endochrome preparing to divide; *c, c*, cells in which division has taken place; *d*, cluster of four cells formed by the repetition of the same process.

161. The cells of the Red Snow are at first of a globular form, and the first step in the process of subdivision is their elongation into an oval shape, and the appearance of a light constriction around them. This constriction shows the tendency of the *endochrome*,—the name by which the colored mass contained in the cell is called—to divide into two parts, each part having a separate envelope; so that the two secondary cells are now included within the external wall of the one primary cell. After this is accomplished, the contiguous

portions of the two subdivisions become so agglutinated as to develop a thick partition between them, by which they become completely divided, and distinct individuals (*b*).

162. While this separation is going on in the interior of the cell and its contents, a thin mucus is being formed on the exterior. In this mucus the cells are embedded; and by the interposition of this new substance, they are liable to being floated to a distance from each other, as shown at *d*, where each of the first pair of the secondary cells has undergone subdivision.

FIG. 10.

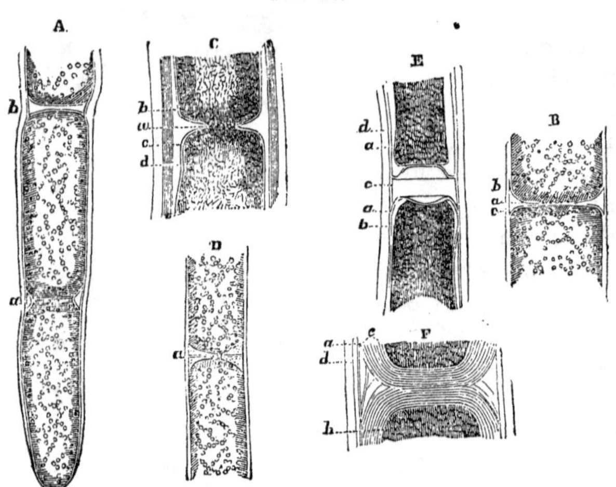

GENERATION OF CELLS BY PLANTS.

A, portion of filament with incomplete partition at *a*, and complete partition at *b*; *C*, more advanced stage; *D*, the separation nearly complete; *E*, the two primordial utricles completely detached, and cell-membrane deposited between them; *F*, successive layers of cell-membrane, making up the thickness of the partition. In the last five figures, *a* indicates the primordial utricle; *b*, the endochrome; *c*, the cell-membrane, and *d*, the mucus investment. A careful study of this figure will enable the student to clearly comprehend all the parts into which the cell is divided.

163. The next degree of development may be illustrated by the Silk Weeds, or Confervas. The Confervas are filamentous plants growing in fresh water. Each filament is composed of a single file of cells adherent to each other, end to end. And the first step in the process of subdivision, consists in the contraction of the primordial utricle or internal investment, forming a sort of hour-glass contraction (*A a*) around the cavity of the primary cell, and ultimately forming a partition (*A b*). The two surfaces of the infolded utricle produce a double layer of permanent cell-membrane between them, and the formation of this may be seen to commence, even before the complete separation of the cavities of the twin cells (*D*). The extension of vegetable structures from cells already in existence, most commonly take place according to this method.

164. In cases where a very rapid production of new and independent cells is requisite, the multiplication is accomplished by the separation of the contents of the primary cell into numerous parts, each one of which parts acquires for itself a covering of cell-membrane. So that a whole brood of secondary cells may thus be at once generated in the cavity of the primary cell, which subsequently bursts and sets them free.

165. This mode of reproduction is also pursued by a species of Confervæ, the *Achlya Prolifera*, which is a parasite plant attaching itself to fish. The Achlya Prolifera consists of a tube-shaped thallus ; the end of this dilates into a cell of large size, and then separates from the other cells by a partition. Within this dilated cell there can be distinguished the circulation of multitudes of granular particles or primary cells. Very soon, however, the endochrome which contains these, separates into a large number of distinct masses,

Fig. 11.

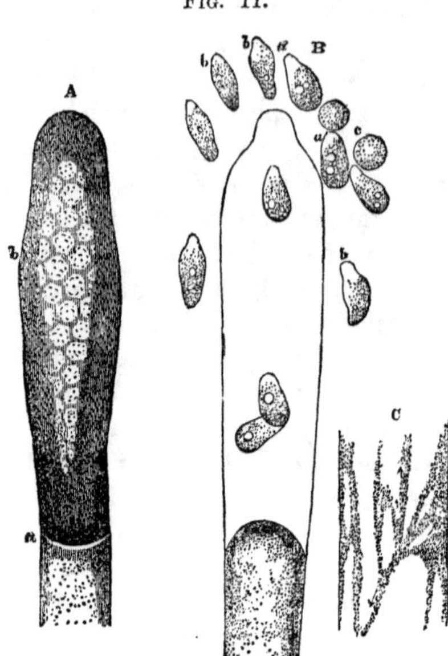

Zoospores of the Protophyta.

A, dilated extremity of a filament *b*, separated from the rest by a partition at *a*, and containing young cells in progress of formation; *B*, conceptacle discharging itself and setting free young cells, *a*, *b*, *c*; *C*, portion of filament showing the course of the circulation of granular contents.

which, at first, are in close contact with each other, and with the walls of the cell (*A*), but they gradually become more isolated, and each one acquires a proper cell-wall; immediately after this they begin to move about within the primary cell. When full grown, these granules are set free by the rupture of the cell-wall (*B*), to go forth and form new attachments on other fish, to acquire for themselves the materials of development into tubiform cells resembling those which gave them birth.

CHAPTER IX.

THE THALLOPHYTES.

1. ALGÆ.

166. The lowest forms of vegetation present no distinction into stem, leaves, or root—no tendency to the formation of distinct organs. They are homogeneous throughout, and the apparent stem, root, and leaves in such plants, exercise none of the functions pertaining to those organs in higher plants. The primordial utricle, by repeated subdivisions, expands into a *thallus*, but the form of this thallus is indefinite, and its whole substance is of the same nature, being constituted of cells of the same kind, without tendency to a distinction of parts. And it is chiefly by a conglomeration of cells, which are necessary to each other, in order to develop masses of the same kind, that there exists in this manifestation of cell-life a distinction from the protophytes.

167. Cells thus grouped together, constitute an order of plants, which are denominated THALLOPHYTES—from *thallus* (leaf-like), and *phyton* (plant). And this order chiefly consists of three tribes of nearly the same degree of development, namely, the ALGÆ, the FUNGI, and the LICHENS.

168. The ALGÆ, are so called from *alga*, which signifies *sea-weed*. They are plants which vegetate exclusively in salt-water. (Fresh water plants are termed *Confervæ*.)

169. The constitution of sea-weeds in general, is very similar to that of the Confervæ. They are leafless plants, with no distinct axis ; consisting either of simple vesicles, of articulated filaments, or of lobed thalli. They vegetate exclusively in water, or in damp situations, requiring no nutriment but such as is supplied by water with the inorganic substances dissolved in it, and air. They absorb food equally by every part of their surface, and they show a great tendency to the extension of the *thallus* by the multiplication of cells in continuity with the existing fabric ; so that sea-weeds frequently attain large dimensions, and sometimes they assume beautiful proportions.

Fig. 12.

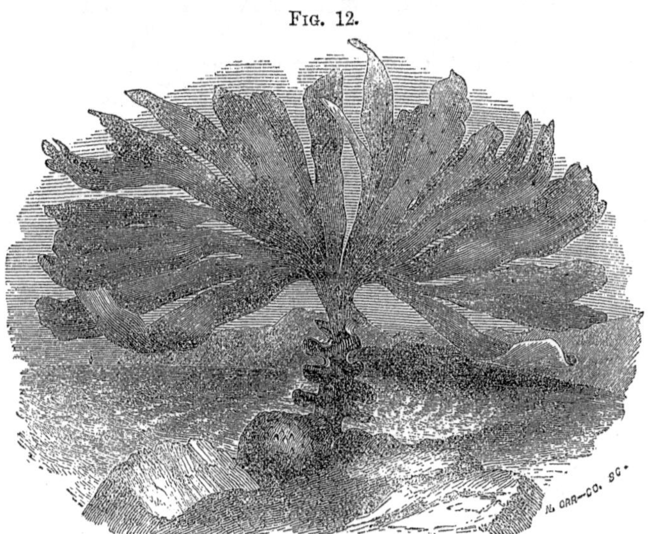

Extension of the Thallus.

170. The *Laminaria Bulbosa*, or what is commonly

called sea-furbelows, sometimes grows to such magnitude, that when fully spread out, a single plant measures ten or twelve feet in diameter. When this plant is young and small, it is supplied with correspondingly frail root fibrils, which are easily detached from the superficial soil of a newly formed rock. It is consequently liable to be repeatedly uprooted, and drifted about from place to place, until it procures a substantial footing. Having obtained this, the radicle swells out into a large hollow bulb, and the external surface of this putting forth new fibrils from every part, prepares, as it were, for the coming storm necessarily to be encountered by the rapid growth and enormous expansion of the thalli. But for this admirable provision at the base, the plant would never be able to stand up against the force of the waves, in the midst of which it surges with a grace and strength comparable with the sturdy oak.

171. Another species of Laminaria, *Digitata*, grows in the deep waters round the Orkney Islands, where it forms an important article of food. In the Orkneys, it is called *Red-ware*, and in some other places, *Tangle*. The Laminaria Digitata, consists of a number of exceedingly strong root-fibrils, which grasp the superficial soil of submarine rocks with great tenacity ; and from this foot-hold, a single, straight, tan-colored stem, as thick as a man's wrist, rises to the height of several feet. The summit of this stem spreads out into a tough palmated thallus, divided into numerous strips, and having very much the appearance of *mace*—if we could imagine a scale of mace to be twelve by eighteen inches in extent. Large quantities of this plant are driven ashore during every storm. And when cooked, it is said to be "good food." A nearly allied species, the *Rhodomenia*

Palmata, is abundant in the shoal sea waters of the British Islands, and is known in Scotland as *dulse*. As thrown ashore by the storms, it consists of thin membranaceous flakes of purplish or redish color. The whole plant is sometimes thrown ashore, and those of largest size rarely exceed a foot in length.

172. This plant is highly nutritious, and cattle feed upon it with avidity. Before the introduction of tobacco into England, it was used pretty much in the same manner as tobacco is now. A few centuries ago, "dulse and tangle," were marketable produce as food for the inhabitants of Great Britain. The same plants are still extensively used by Islanders in various parts of the world

173. In Iceland, *dulse* is gathered and packed into casks with the greatest care. It is first washed in fresh water and dried in the open air, when it becomes covered with a mealy dust, which is sweet and palatable. This dust is deemed the best part of the plant, and in this state it is mixed with rye flour, or boiled in milk and used by the highest class of the inhabitants. In Kamtschatka, dulse is not only extensively used as food, but a fermented liquor is produced from it.

174. The *Iridœ Edulis* is also known as *dulse*, on the southern shores of England. This, however, is a totally different species. It consists of a short stem expanding into an oval thallus sometimes a foot and a half long, and eight or ten inches wide. These, however, are usually made ragged by the force of the waves, and rarely seen entire. They are thick and fleshy, have a smooth, glossy surface, and are of a blood-red hue.

175. HENWARE, of Scotland, is the *Alaria Esculentia* of botanists. On the shores of Ireland, the same plant is called Murlins. This plant is of a transparent yel-

lowish-green color, dries without change, and presents a beautiful appearance. It forms an important article of food for the poorer classes of the inhabitants on the coasts of Scotland and Ireland. An English naturalist, Mr. Johns, gives the following account of its edible qualities : "While walking round the coast near the Giants' Causeway, I once observed a number of men and women busily employed near the water's edge, and on inquiring of my guide, found that they were providing themselves with food for their next meal. Being curious to discover what kind of fare the rocks afforded, I stopped one of the men who was going home with his bundle, and asked him to give me a bit to taste, prepared in the way in which it was generally eaten. He accordingly stripped off all the expanded part of a long and narrow leaf, and presented me with a stem or mid-rib. It was, I must confess, as good as I expected ; but at best a very sorry substitute for a raw carrot, combining with the hardness of the latter the fishy and coppery flavor of an oyster. I made a very slight repast as you may suppose ; and, after having given the man a few pence for his civility, continued my walk. My guide, however, seemed to think that if I did not choose to enjoy to the full the advantage which I had purchased, there was no reason why he should not. He accordingly stayed behind for a minute or two, and when he rejoined me, was loaded with a supply of the same plant, which he continued to munch with much apparent relish as we pursued our walk."

176. A substance has recently been introduced into commerce as a substitue for Iceland Moss. This is "Carrageen Moss," which, notwithstanding its name, is a sea-weed, the *Chondrus Crispus.* When viewed under

water, in its living state, it displays beautiful and ever-changing prismatic hues. As found in the shops it is of very variable appearance, but most usually exists in the form of a thin thallus, of a somewhat triangular shape. It contains a large proportion of gelatin, and has been successfully applied instead of isinglass, in the making of *blancmange* and jellies.

177. One of the most singularly beautiful sea-weeds is the *Padina Pavonia*, or what is commonly known as the Peacock's Tail.

FIG. 13.

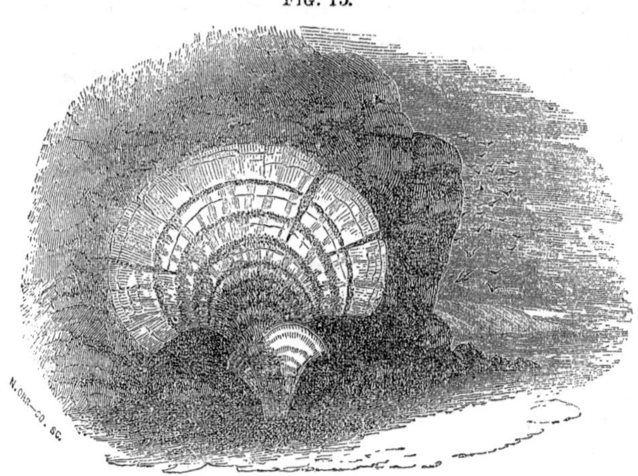

BEAUTY OF THE THALLUS.

178. The *Padina Pavonia* or Peacock's Tail is a common species of sea-weed, generally found attached to rocks in the shoal waters of most sea coasts, and presents a beautiful appearance. It consists of a light olive brown thallus, which springs from a rather firmly fixed rootlet, in the form of a rounded fan, elegantly

marked with concentric bands of a darker hue. The outer margin is delicately fringed with minute filaments, which, as seen under the water in a living state, are in constant motion, and often reflect the tints of the rainbow. One side is generally of lighter color than the other, and has a frosted appearance, as if sprinkled with the powder of pearl.

FIG. 14.

THE BRYOPSIS PLUMOSA.

179. The *Bryopsis Plumosa*, is so named on account of its *plumose* or feathery appearance. It is supplied with strong rootlets and generally found attached to the hard bottoms of shoal water, or on coral reefs. In these situations, it presents a truly beautiful appearance. It is of a bright grass green color, of extreme delicacy, and each branch bears a striking resemblance to a twig of green pine.

180. The *Sargassum Capilliformis*, a species of sea-weed found in the neighborhood of the South Sea Islands, consists of long whip-like stems about one inch in diameter, which give off a great number of delicate tendrils, many of which bear air-vessels about the size and appearance of fully matured Isabella grapes; the whole plant being of a beautiful purple color. It only has rudimentary root-fibrils, and is usually seen adrift, the air vessels buoying it up like so many little balloons; and the tendrils which in this species take the place of thalli, spread out in all directions as if reaching for a place to cling to. And that they may the more easily fasten themselves to the rough rocks with which they are liable to come in contact, they are covered with a glutinous mucus, by means of which they readily adhere to any hard substance.

181. Many other beautiful varieties of sea-weeds there are, a bare catalogue of which would well nigh fill a volume. Indeed, nothing better shows the great purpose this tribe of plants is designed to fulfill in nature's domain, than their immense multitude, enormous growth, and prolific means of reproduction. On the tropical coasts of Africa and South America, fields of sea-weed are sometimes encountered of such strength that ships are turned from their course. And of such unlimited magnitude are some species, that *fragments* of single plants are sometimes seen *several miles* long.

182. In many species, every dissevered fragment is capable of reproducing a perfect plant, while in most, an immense thallus is reproduced in an exceedingly short space of time. Others multiply species in the manner already described in the Achlya, of the Confervæ. Some are provided with reproductive organs

in the form of little capsules called *conceptacles*, usually situated at the apex of the thallus. These contain two kinds of cells designated as germ-cells and sperm-cells. The former absorbs the latter, and in proportion as this process goes on, the germ-cells increase in size like the maturing seed of a flowering plant ; while the sperm-cells diminish and finally disappear. On the disappearance of the sperm-cells, the germ-cells rupture and set free a new set, called embryo-cells, which are virtually young plants. These are distinguishable from the others by being provided with two hair-like appendages, by means of which they are capable of motion, and, therefore, true zoospores. There are yet other species wherein the two kinds of cells are the product of different plants. By these various means, myriads of sea-weeds are created and developed into full grown plants, which furnish the waters with an inexhaustible supply of food for the innumerable animals that dwell therein. And by the ebbing and flowing of the tides, the raging of storms, and other causes, the new rocks are also supplied with food for the nourishment of the plants and animals by which they are inhabited.

2. THE FUNGI.

183. The Fungi require the presence of dead or decomposing matter as a condition of their development. In the language of Lindly, "The Fungi are a provisional creation, waiting to be organized, and then assuming different forms, according to the nature of the corpuscles which penetrate it or are developed among it." For these reasons, some naturalists are of the opinion that the Fungi are equally distant from both the vegetable and the animal organization,—mere

fortuitous devolopments of vegeto-animal matter, called into varied action by specific conditions under circumstances widely different. But the microscope reveals the structure of the Fungi to be very similar to that of sea-weeds, while the conditions of their development, —notwithstanding their alliance to animal matter in chemical composition,—show them to be an order of vegetation higher than those plants before considered, the humus of which is necessary for the growth of Fungi.

184. The Fungi are distinguished for their diffusion and number; for their poisonous properties and their peculiar season and condition of growth; for their exceedingly minute spores, and for their love of darkness. And they have, not without good reason, been supposed to be the cause of numerous fatal epidemics.

185. "Mushroom growth" is proverbial. Some of the smallest species pass their whole existence in a few minutes, and so numerous are they that notwithstanding their short duration, the substances on which they thrive are speedily destroyed. On the coast of Guinea I have had my boots destroyed by them in one night! And some there are, which attain enormous dimensions—to the circumference of seven or eight feet and to the weight of over thirty pounds,—all this in a few days.

186. There are other species which in one night, grow from a small point to the size of the double fist, and the whole of this mass is constituted of cells so small, that such an one contains over *four thousand millions*,—a number which, if counted at the rate of three hundred per minute, it would take one person *night and day three hundred years to count.* A square mile contains over *three millions* square yards or *twenty-*

seven millions square feet, so that a single fungus may present at evening an almost invisible single cell, and yet before morning place nearly *eighteen hundred* such cells in *every square foot* of a square mile!

187. The number of *species* in this tribe of plants is no less remarkable than the number of individuals. Fries, a Swedish naturalist, observed no less than *two thousand species* within the compass of a square furlong.

188. The Fungi possess the curious property of destroying their own reproductive powers, or of poisoning against themselves the soil in which they grow. Persons who have traveled much over our western prairie lands, cannot have failed to observe curious denude circles, amidst the vivid green of rank vegetation. Such spots used to be attributed to the tiny feet of fairies, who were supposed to make the spots so marked, their place of revelry, hence they were called *Fairy Rings.* Others attributed them to the revelries of a worse sort of spirits, the witches. Subsequently they were thought to be the effect of electrical action.

189. Fairy Rings are now known to be produced by the eccentric growth of various species of fungi,—they may, therefore, be termed the vegetable ring-worms of the fields. Commencing as animal ring-worms at a small point, these fungi move progressively outwards, leaving a bare uninviting space behind them, upon which, for a time, neither other fungi nor grass will grow. Finally, however, grass returns, its seeds take root first at the centre, which part has been longest barren, and filling this up, follows the course of the fungi so as to produce a broad circular belt of scorched earth, which expands more and more in diameter. The fungi, evolved only on the outer edge of the belt, do

not again attack the centre, in which the soil seems to have lost its power of sustaining them.

190. The species of fungi, commonly called mildew or mould, is so small as to be invisible, except when collected together in large numbers, and they are so light as to readily float in the atmosphere, and are thus subject to being inhaled. On being planted by this means into the blood of the animal system, the conditions are congenial to their multiplication.

191. Several years ago, after being some weeks at sea, last from the Gulf of Maracaybo, one of my messmates for the first time in his life was taken sick with intermittent fever. On the first morning of his illness he complained of a musty odor about his bedding, which lead to breaking out his clothes-lockers underneath his bunk. In these lockers he had stowed various curiosities,—carved cocoanut shells, boxes and bags made of seeds, rare nuts, etc., purchased of the natives. All of these things were found to be mouldy. Previous acquaintance with the habitudes of the fungi, had led me to suspect them as the cause of intermittent fever and other diseases, but up to this time I had no positive proof.

192. The lockers and contents were thoroughly cleansed with boiling-hot salt water, and every thing mouldy purified or discarded. A few days afterwards, two others of only six in the ward-room, were also taken sick with intermittent fever. And here it may be remarked that up to this time, there had not been a case of this disease on board during our eight months cruise on the Central American coasts. Under my direction we now had a general over-hauling and breaking out of every thing in the apartment. Adjoining the locker first found to contain mould, was the mess-

locker, filled with edible stores. Here we found Pecan and Hickory nuts mouldy *inside.* By this circumstance I was struck with the idea of the exceeding smallness of the sporules of the fungi, for how otherwise could they get *into* the nuts? The novelty of the idea induced me to lay aside some of the suspected nuts for particular examination, and for this purpose I carefully placed them in a drawer under my bunk. The rest of the nuts were thrown overboard. The next day, complacently enjoying the cleanly state of our quarters, now sweeter than a nut, I cracked one of those laid up for examination. Compared with those looked into the day before, there was an increase of opacity or blackness of the mould, and evident progress in the destruction of the kernel. The second day, progress was still more marked, and on first breaking the hull, there was an escape of a smoke-like dust, which seemed for the most part to consist of the contents of the nut now reduced to that state. On the third day, I found the nuts cracking open, they were riven by the infinitely small fungi! I at once had them all thrown overboard. Less than a week afterwards, I was rewarded for my investigation by an attack of intermittent, the which my mess-mates enjoyed exceedingly. Subsequent service on the coast of Africa and other notoriously unhealthy places, afforded me additional opportunities for investigating the nature of fungi, and I have long since concluded that this group of plants often causes other and worse diseases than intermittent fever.

193. During the pestilential season in New York, in 1798, Webster informs us that he saw a cotton garment covered with dark grey colored spots of mildew *in a single night,* and that three years before, *sound*

potatoes perished in his cellar *in thirty-six hours*, and that such events were common.

194. Dampness, darkness, and a temperature about the same as that of human blood, are the conditions under which fungi seem most to flourish. Water is supposed to act unfavorably to the production of fungi, by means of its coldness. And the practice of hydropathy shows the little hazard of wet sheets. But experience demonstrates the danger of *damp* and *musty* bed-clothes, hence a limit of the application of water should be constantly kept in view.

195. The power of salt in the prevention of the growth of fungi, is abundantly proven by the habits of sea-faring men. Sailors who have been exposed to drenching rain, and who are frequently obliged to sleep in the open air, commonly practice redrenching themselves in salt water, to "keep from taking cold;" it is in effect an antidote to the growth of fungi in their clothing.

196. Another means of partially depriving fungi of their usually poisonous qualities, is that of *drying* them. This process, like that of the salt water drenchings, practiced by mariners, is frequently resorted to when the persons who practice it know nothing of the poison they act upon. The healthy effect of a "*fire in the camp*" in tropical marshes is known to every experienced soldier, but *why* it has this effect, they are generally ignorant. It is because the fungi which most abound in the night, and above all in tropical latitudes, are *dried*, and, in a measure, rendered innocuous.

197. Folchi, a Roman writer, says, "Many persons are known to me who have during many years preserved themselves from fever, in the worst parts of the country around Rome, by adopting the most rigid cau-

tion in retiring within their houses before evening, closing the windows, *warming the rooms*, and taking care not to go out in the morning until the sun has been some time above the horizon."

198. Like man, the fungi live in any climate, and although the poisonous varieties are most common to hot, damp climates and places, no climate nor place is wanting in the conditions which give rise to some species. It is a peculiarity of their nature, that even the same species is rendered dangerous in proportion to the conditions which most promote their growth. And persons who close their houses during the Summer, frequently return to them in the Autumn to be made sick by the fungi. To avoid this, it would be well to remember the soldier's caution, first "build a fire in the camp."

199. Another curious propensity of the fungi is, to

A FIG. 15. B

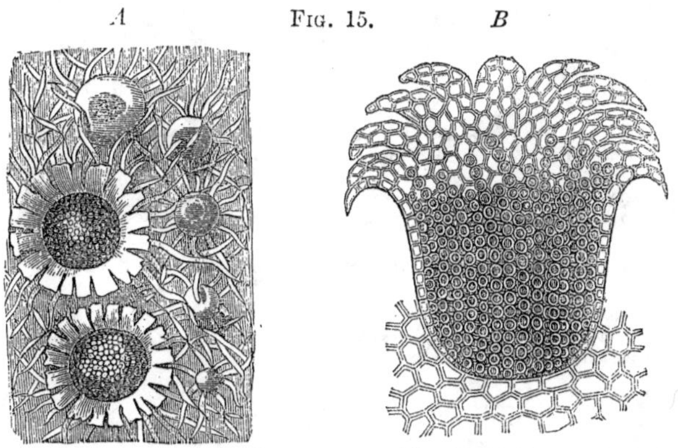

STRUCTURE OF THE FUNGI.

A, Magnified portion of *Æcidium* tussilaginis ; *B*, section of one of the conceptacles with its spore cells.

infest neglected books and manuscripts. The nature

4

of the material of which books are made, renders them exceedingly liable to become mouldy, and this mould if inhaled is irritating in its effects ; so that individuals frequently contract severe "influenzas," and sometimes worse diseases, by overhauling a too long neglected library. In structure, the fungi consist of elongated branching cells variously interlaced with each other, but unconnected and capable of separate existence. But this arrangement of the cells is so exceedingly various, that from the thallus alone the species cannot be determined. And any portion of the thallus which may be separated from the rest, is capable of reproducing the species in immensely numerous spores, which are so light as to float in the atmosphere, and being borne on the "wings of the wind," they are deposited in places the most remote from each other, and give rise to vast numbers for new generations.

FIG. 16.

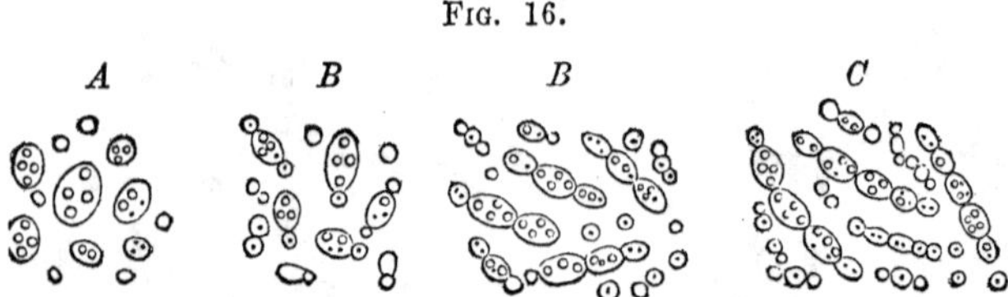

AN EDIBLE SPECIES OF FUNGI, THE TORULA CEREVISIŒ OR YEAST PLANT.

A, fresh yeast—single cells of which it at first consists ; *B, B*, yeast in a lump for one hour, cells with buds ; *C*, yeast in a lump for eight hours, cells united by continuance of the budding process.

The Yeast Plant at first consists of transparent cells overlapping each other. When cells like these are placed in a lump, no matter how large, they are the little leaven that leaveneth the whole.

200. And this is the course they pursue :—Immedi-

ately that they are put in the lump, or supplied with a *pabulum*, they begin subdividing themselves by a budding process, and after about an hour's growth, they have changed in appearance, as represented at *B*. In *eight* hours, or by morning, when the yeast has been added to the lump over night, and when the dough is ready to be made into cakes, the cells have arranged themselves into lines, and begin to form myriads of simple transparent cells, like the first. Thus it is that the quantity of yeast first introduced into any fermentable substance is many times multiplied during the process of fermentation. And if this process be permitted to continue, each of the necklace like filaments extends itself by the production of new cells ; but if, at any time, the fermenting process is checked, the cells separate and assume the condition of those which originally constituted the yeast.

201. The edible *tufas* of the Neapolitans, which are, by some persons, esteemed a great delicacy, are produced in shaded excavations on volcanic soils, or in caves and in cellars, where the conditions for the growth of fungi most exist. And in the Maremma, where the *tufa* is the basis of the soil, the surface is intermixed with the animal remains of departed empires, where everything conduces to a *fungiferous* condition.

202. In the vicinity of Rome, the soil is so prolific in organic remains, that the *edible* mushrooms produced, are a source of considerable revenue to the Papal government ; over ten thousand dollars worth being taxed annually. The quantity of mushrooms yearly sold in Rome, has been estimated by weight to be not less than 140,000 pounds.

203. *Mushroom-Stones*, which are exported from Italy,

are a species of volcanic tufa, very porous, and of an argillo-calcareous nature. The pores of the stones are filled with the soil from which they are taken. "And" (in the words of the English Philosopher Boyle, who first described this stone, under the title of *Lapis Lyncurias*) "which, when moistened and warmed, will in a very short time produce mushrooms fit to be eaten."

204. Considered as an article of diet, *Mushrooms* and *truffles* are at least very uncertain in their qualities. Some species are reckoned exceedingly delicious and nutritive, but all are hard of digestion. Besides, great care is at all times necessary, in the selection of edible species, as they are generally associated with many that are injurious, and some that are virulently poisonous.

205. Most *fungi* which have a warty cap, and those which have a membrane adhering to their upper surface, are poisonous. Heavy fungi, which have a disagreeable odor, especially if they emerge from a vulva or bag, are hurtful. Of those usually found growing in woods and shady places, a few are esculent; but the greater part are unwholesome, and if they are moist on the surface, they are generally poisonous. All those that grow in tufts or clusters from the stumps or trunks of trees, should be rejected. Wherever collected and whatever the appearance, if a mushroom have an unpleasant pungent odor and astringent styptic taste, it should be rejected as poisonous. All those which are corky in texture, or have a membranous collar round the stem, or which, on being cut, soon after become bluish or turn red, are poisonous.

206. Finally, with the greatest care in the choice of mushrooms as an article of food, accidental poisoning from the use of deleterious varieties is by no means

uncommon, and the habitual use of them under the most favorable circumstances, is frequently followed by diseases, of which they are known to be the cause.

207. The free use of *vinegar* in dressing mushrooms, is an excellent precaution against the ill-effects of some species, otherwise poisonous ;—vinegar having the effect of partially neutralizing the acrid principle. And it is the custom of some persons never to eat mushrooms without vinegar.

3. THE LICHENS.

208. Next to the fungi in the vegetable world, come the LICHENS. These vegetate upon rocks, walls, trunks of trees and other wood, upon hard earth or upon other things sparingly supplied with moisture, provided that there is free access of air and light.

209. The lichens may be recognized in the form of simple dust, crispy membranes or leaf-like appendages to branching stems. (Plate II.)

210. In constitution, some orders of the lichens resemble the sea-weeds, and others resemble the fungi. But there are other species more highly organized than either the sea-weeds or the fungi. The general tendency of development in the lichens is to form a hard crispy thallus of slow growth and great durability.

211. The upper surface of the thallus being most exposed to light and air, becomes hard and dry, a condition which favors the evolution of the fruit ; whilst the under surface is soft and adapted to the absorbtion of nourishment. The under surface also frequently in part consists of root-like appendages, which not only serve to fix the plant, but aid in the absorbtion of moisture for its nourishment. These appendages, how-

ever, are of the same structure as the thalli, and the only analogy they bear to true roots, is that of supporting the plant in an erect posture ; these plants being nourished by absorbtion from the atmosphere.

212. The higher orders of lichens multiply species in a manner similar to the higher orders of sea-weeds, by means of two kinds of cells, contained in *conceptacles.* When the *spores* or embryo-cells are produced, they are easily detached by the winds, or washed out by rain, and being extremely light, they are carried great distances in the air. In some species, the spore-cells are colored and bear strong resemblance to the seeds of higher plants.

213. The lower orders multiply by subdivision and budding, in the same way as the yeast plant and other species of fungi. The reproductive cells being situated between the layers which compose the upper and the under surfaces of the thallus, the cells being aggregated together in little green masses of various sizes. These masses gradually find their way through the upper or cortical layer and appear on the surface as little dust-like patches, each particle of which, when dispersed by the wind, is capable of developing itself into an entire plant.

214. In composition, the lichens consist in part of a starchy substance that is highly nutritious. And in the Arctic regions where other vegetation is scant, the lichens are of great importance. When wet, they appear like green herbage, and in this state they are tender, and some animals greedily partake of them. A species grows in Lapland, termed Reindeer-moss, which furnishes almost the sole food of that useful animal to the Laplander.

215. One species, the *lichen icelandica,* commonly

known as "Iceland-moss," has often formed an important article of diet for man. It is chiefly produced in Norway and Iceland, and great quantities of it are annually exported to other countries; over 20,000 pounds have been sent to England alone, some years, and it is justly deemed an important article of stores for all Arctic voyagers.

216. A jelly prepared from "Iceland-moss," possesses immemorial reputation as an article of diet for the sick;—in preparing it, the bitter principle of the lichen should be, *in part*, removed. This may be done by steeping it in a large proportion of lukewarm water, which may be rendered more efficacious by the addition of a little potash. Some of the bitterness, however, should be retained, as to this is due the tonic virtues of the plant. The jelly properly prepared, is nutritious, easily digested, and well adapted to the delicate stomach.

217. The "*Tripe de Roche*" (*Gyrophora proboscidia* and *G. Cylindrica),* are scarcely less celebrated as articles of food in the Arctic regions, than the *lichen icelandica.*

218. Other species of lichen, *Dyers Orchil, Litmus, etc.,* furnish important coloring principles, and are extensively used in the arts for dying and chemical purposes.

CHAPTER X.

THE ACROGENS.

1. THE MUSCI OR MOSSES.

219. The second grade of organic development consists of plants, which, as their stems usually grow by elongation from the apex, and have little or no provision for increase in diameter, have been termed ACROGENS, from *akros* (point), and *gigno* (I grow), that is, *point-growers*. The THALLOPHYTES, for the sake of uniformity, may be called THALLOGENS.

220. In this type of organization, as in that of the Thallophytes or Thallogens, there are three tribes of plants which chiefly constitute it. These are the *Musci or Mosses*, the *Hepaticæ or Liverworts*, and the *Filices or Ferns.*

221. The mosses are commonly understood to be small flowerless plants, of but little interest and of "no known use" (*Brande*). But the economy of nature admits of no such distinction. In Spitsbergen, the rocks which rise from the surrounding ice are thickly covered with moss. In other Arctic regions and extremely cold climates where all other vegetation—or all but that of the humble tribes already considered—fails, the surface of the deeply frozen earth and ice-cold rocks is clothed in rich green moss. And even in our own climate, the coldest winters are characterized by a corresponding luxuriance in the growth of the mosses.

222. In all this there is Infinite wisdom and utility. The mosses shelter and preserve the germs, seeds, roots, and embryo plants of many vegetables which would otherwise perish. They also furnish material for both food and shelter to insects, birds, and other animals, which in their turn furnish food for man!

223. Nor is this all. When the Summer's sun returns, the mosses wither and die, and in *their* death and decay, food is supplied to a rising generation of plants. Other mosses grow in bogs and marshes, and by continued multiplication and decay, fill up and convert them into fertile pastures, or into peat-bogs, the source of inexhaustible fuel to the inhabitants of polar regions.

224. The organization of mosses, is no less interesting and beautiful than the purpose they serve in the economy of nature. They always have a distinct axis of growth (Plate II.), a more or less erect form, and incipient leaves symmetrically arranged.

225. A transverse section of this axis displays a variable structure between the outer and inner portions, by the intervention of a layer of elongated cells, that seem to prefigure the structure of wood in the higher plants. From this layer, prolongations of condensed cells pass into the leaves, forming *mid-ribs.*

226. The structure of the leaves, however, do not present much advance towards a more perfect type of plants. The leaves being merely a solidified aggregation of cells.

227. Root-fibrils are observed to put forth from the lower portion of the axis, and small fibrils of the same character exist on the under surface of the leaves, but they do not possess the functions of roots. The conical summits (Plate II.), are *encasements* filled with sperm-

cells. And on various parts of the surface of the thallus are other little cases, denominated *thecas;* these are basket-like conceptacles, one of which is here represented, largely magnified. Within this conceptacle, there are a number of spore-cells. These in the *theca* are covered over by a thin membranous hood or lid, called *operculum.* The inner surface of this lid is covered on the edges by rows of little bristles, which are, collectively termed, the *peristome*, a word which means "around the mouth," where the bristles seem to be placed for the purpose of admitting air and light to the contained spores ; and, besides this, to afford entrance to the *fovilla,* the name given to the mucus containing sperm-cells, in the encasement. These sacs are analogous to the stamens in flowering plants, and on the application of water to them as rain, they discharge the fovilla, which is filled with germ-cells, and these falling upon the lid of the conceptacle, are admitted through its open work, and brought in contact with the germ-cells therein contained. By this union of the sperm and germ-cells in the conceptacle, the *embryo or spore-cell* is produced, and, becoming a zoospore, goes forth to develop itself into a new plant.

Fig. 17.

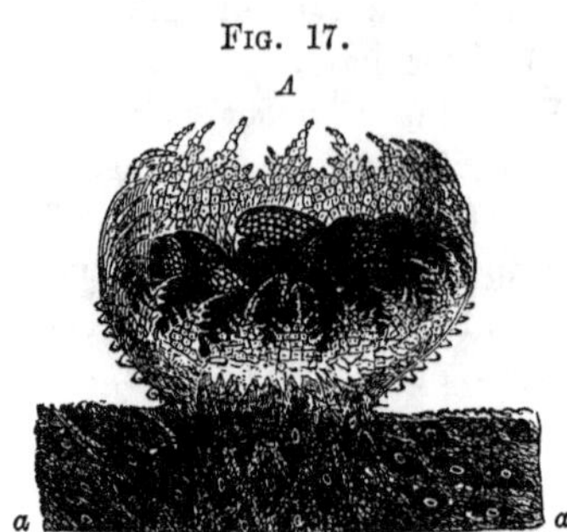

Conceptacle of Mosses and Liverworts.

A, conceptacle fully expanded, rising from the surface of the frond, *a*, *a*.

2. THE HEPATICÆ OR LIVERWORTS.

228. The Hepaticæ or Liverworts were so named by the ancients, on account of the supposed virtues of one species, the *Marchanta polymorpha* (Plate II.), in combating the diseases of the liver.

229. The constitution and habits of the Hepaticæ, strongly comport with those of the Lichens, but their manner of reproduction places them higher in the scale of organization than even the Mosses. For, in the Hepaticæ we have the first clear distinction of species into sexes ; that is, the sperm-cells and the germ-cells are born upon different plants. A few of the Mosses also have this distinction. The male plants are provided with broad *peltate* or shield-shaped receptacles, bearing *antheridia* or membranous sacks analogous to the stamens of higher plants, containing sperm-cells ; and the female plants have lobed receptacles containing *pistillidia* or germ-cells. Neither of these plants is capable of multiplying its species,—the sperm cell of the one being necessary for the development of the germ-cell of the other into the embryo or spore cell. The *theca* or spore capsule of the thallus in the Liverworts, is entire, that is to say, it has not an open work lid like the theca of the mosses, but instead of this, it cracks open into two or four equal parts.

230. In the process of development, the thallus of the Liverworts presents various gradations of form, from that of a simple amorphous expansion to an assemblage of regularly arranged parts ; and from every part of the under surface, root fibrils are given off, but these remain in the same simple condition as when first put forth—their texture being in all respects the same as the thallus.

3. THE FILICES OR FERNS.

231. The lowest orders of Ferns (Plate II.), show but a slight degree of elevation above the Liverworts. Many of them when full-grown only measure a few inches in length, and several species are parasites and grow upon the bodies of trees.

232. The ascending axis of the simplest ferns, is in its structure similar to that of the Mosses and Liverworts, combining the leaf and stem, and therefore only a variety of the thallus. But this organ in the Ferns has been more particularly distinguished by the name of *frond*, a word frequently used synonymously with thallus, and defined by Linnæus to be a stem "in which the leaves are confounded with the stem and branches, and frequently with the flower and fruit." Linnæus restricted the term *frond* to Ferns and Palms.

233. The *fronds* of ferns are entire, variously divided and veiny. From the fibro-vascular bundles in the stem, prolongations pass into the leaf-stalks, and thence into the mid-ribs and lateral branches of the frond, to which the condensed layers form a kind of skeleton. They may therefore be said to consist of a *cellular* parenchyma or *pith*, inclosed between two layers of thin bark called *epidermis* (from *epi*, upon, and *derma*, the skin), through which there are numerous external openings called *stomata* or mouths.—In the use of the word *cellular*, its meaning should be carefully distinguished from that of cell-formation. Cellular, cellular tissues, etc., means having *spaces* or cavities between the aggregation of cells of which the substance is formed; these cavities usually contain air.

234. The following figure represents the process by which Ferns are multiplied. The *thecæ* or sporangii,

FIG. 18.

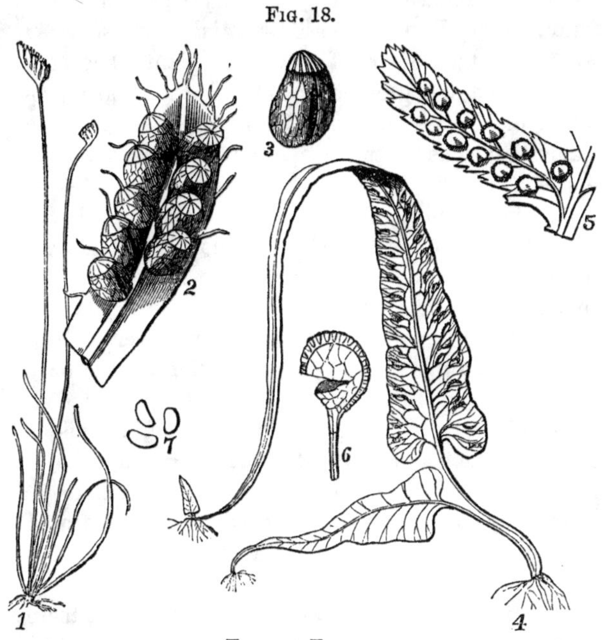

FILICES—Ferns.

1, Schizæa pusilla of about the natural size. 2, A section of a fertile frond, magnified, showing the thecæ or sporangii, occupying the under surface. 3, One of the thecæ more magnified, and opening by a cleft. 4, Asplenium chirophyllum or Walking Fern; the fronds rooting, as they frequently do, at the apex. 5, Section of a frond of Aspidium goldianum, the sori attached to the veins. 6, Magnified thecæ or cell-case of Aspidium, with its stalk and elastic ring partly surrounding it, which tending to strengthen itself when dry, tears open the thecæ, and sheds out the minute spores, 7.

are capsules containing spore-cells, and the *sori* are clusters of thecæ or sporangii. The spore-cells of Ferns evolve both sperm and germ-cells, and these, by contact with each other, produce new thecæ filled with embryo-cells, or young plants.

235. This apparently complicated process of reproduction in the Ferns, though essentially the same as that of the other orders of flowerless plants, shows a tendency to a new grade of development, in the reproduction of species by the different parts of a single organism such as takes place in flowering plants from the seed.

236. The root-fibrils in these simple varieties of Ferns are similar to those of the tribes previously considered, partaking in a great measure of the same constitutional character as the thallus or frond.

237. Among the more elevated species of ferns, with which we are all more or less familiar, there is not only an ascending axis, but around it foliaceous appendages symmetrically arranged, and a descending axis or root, from which alone radical fibres are given off.

238. In these higher varieties, the stem consists of separate layers,—an outside *cortical* portion separated from the inner portion or *pith* by the interposition of a layer of condensed cells, called *woody-fibre.*

239. It is only in the Tree-Ferns, however, that there is a perfect evolution of the character of this tribe of plants. A species of Edible Fern found in New Zealand, grows to the height of twenty feet, and has a stype or body four feet in circumference. The under surface of the fronds of this species are of bright pink color. It grows most luxuriantly in shady places, from deep crevices in the rocks. But sometimes, by the removal of other growth, it is left standing on naked precipices, and in these situations it presents a most magnificent appearance.

240. The number of species of Ferns known to Botanists amounts to nearly one thousand. They mostly abound in moist, shady places on sea-coasts and newly

Fig. 19.

Cyathea Dealbata—Edible Fern.

formed islands, in tropical climates. And in some countries certain species serve for aliment to the inhabitants. But the Edible Ferns are gross and unsubstantial food, only suited to the still grosser nature of the wretched inhabitants of New Zealand, New Holland, and similar climates where such ferns grow.

241. The *Filix mas*, *Polypodium calaguala*, *Osmunda*

regalis, and a few other species have been used for medicinal purposes, but their properties are uncertain, and they are but little esteemed.

242. The great purpose which the ferns answer in nature is that already contemplated in Chapter VI., namely, to furnish material for succeeding generations of plants.

243. Whatever diversities there may be among the plants now considered, these diversities are attributable to the deficiency or excess of some or other of the component parts which enter into their organization as a whole, and the same analogy exists in the higher forms of organization. Many trees, the California pines, for instance, are remarkable for the great size of their bodies and stems, while they are deficient in the development of the essential parts of the flower. Many small plants, on the other hand, are remarkable for the beauty and luxuriance of their blossoms, but deficient in a woody stem. The difference between these conditions does not consist in the elevation of one above another, but in the diversity of plan on which the same elementary parts are combined and arranged.

244. Considered in relation to the vital functions, the simple plants differ from the most complex chiefly in this : that the whole external surface of the *Cryptogamic* or Flowerless Plant operates equally in all the functions of life, as those of Nutrition, Respiration, Reproduction, Exhalation, etc. ; whilst in the *Phanerogamic* or Flowering Plant, these functions are confined to *certain portions* of the structure. Cryptogamic, is derived from two Greek words signifying concealed marriage, and as applied to plants, means plants which have stamens and pistils concealed, or more properly concealed flowers. In the orders of plants above con-

sidered, the stems, roots, leaves, and flowers, are so combined together in the thallus or frond, as to have all the functions performed, as by one organ ; they have neither flowers nor seeds, but are propagated by *spores*, bodies analogous to seeds. Hence, such plants are distinguished by the name of FLOWERLESS PLANTS, or Cryptogamia. While plants which have organs distinguished by the performance of special functions are called FLOWERING PLANTS or Phanerogamia, from *phaneros* (evident), and *gamos* (marriage).

245. As we proceed from the lower to the higher forms of organized structures, we shall perceive that *progression from the general to the special attributes of organic bodies,* is a general law of organization.

246. If we compare the forms which the same structure presents in different species of the same tribe of any order of beings, we find that the structure exists in its most general or diffused form in the lowest species; in its most special and restricted form in the highest species ; and that the transition from the one to the other species, is always gradual. The function, therefore, which is at first most general, and is so combined with other functions as scarcely to be distinguishable from them, is afterwards found to be limited to a single organ, or to be *specialized* by separation from the rest. All the other functions are appropriated in the same manner. Hence, the whole work of the organism comes to be accomplished by a "division of labor" amongst its several organs, each of which is adapted to execute its particular share with a measure of energy and completeness proportionate to the specialty of its development.

CHAPTER XI.

FLOWERING PLANTS—PHANEROGAMIA.

247. The term flowering plant does not necessarily indicate the possession of what are commonly termed "flowers," because these, in the lower orders of the series of plants now about to be considered, exist in such a rudimentary state as to be scarcely distinguishable from the thecæ and spores of the Cryptogamia. The distinctive character of the Phanerogamia or flowering plants, consists in their reproduction by SEED;—that is to say, by means of an organic body which is the matured result of fecundation, and designed to reproduce the species. A SEED comprehends in miniature all the organs essential to the constitution of an adult plant, and it is only by the process of germination that the organs of a flowering plant are successively developed.

Fig. 20.

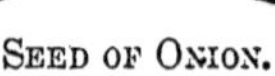

SEED OF ONION.

a, *a*, Albumen ; *b*, Embryo.

248. The seed may be considered as that link in the chain of vegetable existence which connects the old

and the new plant. It consists of a *nucleus* usually enclosed in two *integuments.* The outer one of these integuments is called *testa* or *episperm,* and varies much in texture and form. It is sometimes furnished with a tuft of hairs at one end, called a *coma;* or is covered with a long wool. The inner integument is termed *tegmen.* The scar left by the separation of the seed from its stalk is termed the *hilum.*

249. The *nucleus* or kernel consists of the *albumen,* and the *embryo.* The albumen is a mucilaginous or starchy substance contained in the cellular tissue of the nucleus, for the protection and nourishment of the embryo. In most seeds, the albumen is of pulpy consistence, but in the grasses—as wheat, rice, etc., it is *mealy;* in coffee, it is *horney;* in the poppy, *oily,* etc. Where the albumen is not deposited uniformly, the nucleus has a wrinkled or folded appearance, and it is said to be *ruminated.* But in some seeds the albumen is wanting, such are called *exalbuminous,* to distinguish them from the larger class, where it is present, termed *albuminous* seeds.

250. The *embryo* is the most important part of the seed ; indeed, all other parts seem subservient to this, because it is the point from whence the life and organization of the future plant originate. The embryo is usually central and enclosed by the nutritious portion of the nucleus ; sometimes it is no more than a mere point or dot, and in some cases it is wholly invisible to the unaided eye.

251. The embryo plant consists, first of the *plumule* and *radicle,* an ascending and descending portion ; second, the *amnios,* a thin membrane surrounding the plume and radicle. During the development of the plume and radicle, the amnios is generally absorbed,

but sometimes it remains and is then called *vitellus.* In exalbuminous seeds, the whole substance of the nucleus is absorbed and the embryo occupies its place. Next to the amnios, are the *Cotyledons* (from the Greek word *Kotule,* a cavity). These are the thick fleshy lobes of seeds which first grow out of the ground, in the form of two large leaves. The cotyledons are the first visible leaves in all seeds, often fleshy and spongy, and consist of a succulent and nourishing substance, which serves for the food of the embryo at the time of its germinating. But after a time, when the young plant becomes sufficiently vigorous to sustain life by means of the ascending and descending portions, the cotyledons usually wither and die. Their number varies in different plants. In the cryptogamia there are none, hence these are frequently classed by botanists as ACOTYLEDONOUS plants.

Fig. 21.

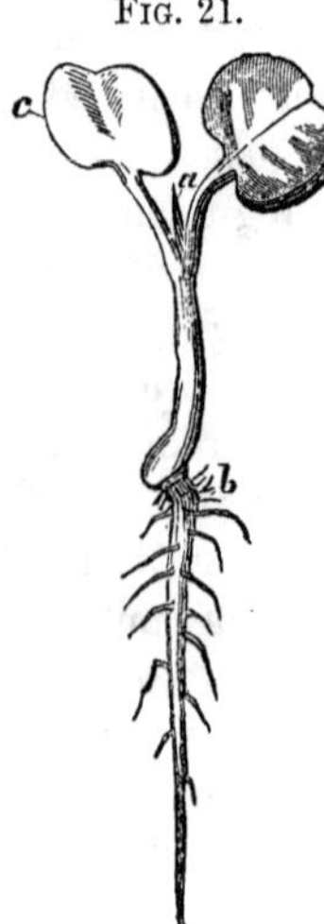

THE EMBRYO PLANT.

a, plumule ; *b*, radicle ; *c*, cotyledons.

252. MONO-COTYLEDONOUS PLANTS are such as have but one cotyledon in the seed ; as the *grasses, liliaceous* plants, *palms, etc.*

253. DI-COTYLEDONOUS PLANTS are such as have two cotyledons ; they include the greatest proportion ; as the *leguminous* seeds—peas, beans, etc.

254. POLY-COTYLEDONOUS PLANTS are those of which the seeds have more than two cotyledons ; the number of these is small ; *hemlock* and *pine* are examples.

255. When a mature SEED is detached from a plant,

it becomes a self-sustaining structure, and in it exists all the rudiments of the plant that produces it, together with a supply of nutriment for its future development. But the principle of life contained in the seed remains dormant, until placed in circumstances favorable to vegetation. When thus placed, it shoots forth and the process through which the seed passes in the first stages of development is termed *germination.* How long seed may lie dormant, is shown by the fact that grains of wheat found in a Mummy case, which must, therefore, have been at least 3,000 years old, have been known to vegetate and produce healthy plants.

256. The conditions requisite for germination are, *moisture,* the *presence of air,* the *absence of light,* and *warmth.*

257. Moisture acts in several ways ;—it softens the integuments, pervades and softens the nutritive matters, and thus brings them to a fit state to be absorbed by the embryo ; conveys in solution nutritive particles from other sources, and by its own decomposition, affords two of the most important ingredients of vegetable bodies.

258. The presence of air is necessary, because the oxygen contained in it causes a change in the starch comprised in the albumen or cotyledons and converts it into a semi-fluid substance, consisting of sugar and gum, and combines with some of the carbon, forming carbonic acid, which escapes, whilst the proportion of oxygen and hydrogen are augmented.

259. The absence of light is favorable, as its presence has a tendency to produce an opposite change, the accumulation of carbon.

260. Warmth promotes the necessary chemical changes, and assists the moisture in acting upon the

hard parts of the seed, and also acts as a stimulus to the absorbents of the embryo. The degree of heat required, is much the same in similar species, but varies greatly in different plants. Thus the seeds of some plants will germinate at a temperature near that of 32°, whilst those of others require a heat of 90° to 110°.

261. Seeds germinate most freely in Spring and Summer, as at these periods the requisites for this process are all afforded to the fullest extent. This is also favored by covering them loosely with soil, so that whilst the light is excluded, they may experience the vivifying influence of the sun's rays, and at the same time be kept in a moist state. When planted at too great a depth, they remain torpid, from not receiving the stimulus of air ; and when they have not a proper covering of earth, they do not germinate, from not obtaining sufficient moisture.

262. When germination commences, the moisture absorbed softens all parts of the seed ; a chemical change takes place in the starch of the cotyledons ; the embryo enlarges and bursts the integuments ; the radicle protrudes and descends, often attaining a considerable length before the plumule appears ; the albumen is gradually absorbed ; the cotyledons expand and become seminal leaves which afford protection and nourishment to the infant plant, by elaborating sap; and when the true leaves appear, the cotyledons are gradually absorbed and disappear, or wither and remain at the foot of the stem. As soon as the root and leaves are developed, in opposite directions, the process of germination is perfected, and the new plant is formed.

263. If we analyze the fabric of a flowering plant, we

find it to consist of an *axis* and *appendages ;* the former being made up of an ascending portion or *stem,* and of a descending portion or *root,* with their respective ramifications.

264. THE ROOT is that part which descends into the earth, acts as a support to the plant, and absorbs nourishment from the soil for its sustenance. Roots, however, are not always produced under the ground ; they sometimes arise from any portion of the stem ; as the branches of woody plants when bent down and covered with earth, will put forth roots. This takes place naturally in many trees in tropical climates, which give off root-branches, high above the earth. Such is strikingly the case in the famous Banyan tree. Roots also attach themselves to the stems of other vegetables ; plants of this character are called *parasites.* Roots increase in length chiefly by the addition of new matter to their extremities. It is by these extremities, called *spongioles,* that absorbtion takes place. The power which roots possess of continually adding new living matter to their extremities, enables them to penetrate the solid earth in which they grow, to insinuate themselves into minute crevices, and to extend from place to place, as the nourishment in their immediate vicinity is consumed.

265. The addition to the ends of roots is always made in the direction of least resistance, and this is the reason why the pervious spots in firm soils and the crevices in walls and rocks are penetrated by slender roots.

266. And when these filaments have once insinuated themselves, by means of the fluid they absorb, they continue to extend and to expand, and it is thus that the openings they have penetrated become enlarged,

and in some cases to such an extent as to rupture masses of rock.

267. By the same tendency—towards points of least resistance—roots grow towards moist places, which are softer than others. By keeping this circumstance in view, the so-called instinct of plants is understood, and may be made practically useful. It is well known to agriculturists that the course of large drains is liable to be interrupted by the extension of roots, not only from trees but also from apparently insignificant plants.

268. Roots are, in some cases, directed by the dispersion of watery vapor through the atmosphere. I have noted an instance of a willow standing fifteen feet from a brick well, where a single root ran in a direct line for a point of the wall of the well, where there was a small aparture left for a step, by the omission of a brick. This step-hole was several feet from the surface of the water in the well, and the root having passed through the hole, on the ground of instinct as applied to animals, we may suppose was at a loss to know which way to turn, but not so with the root; having passed through the aperture, the vapor of the water directed its course. After first dividing itself into a tuft of new fibrils, so soft as to be capable of nourishment by the absorption of vapor, these fibrils descended and formed a large mass at the bottom of the well, eventually destroying it. By the influence of vapor, roots also divide over a stone, which may stand in the way. If, as in the case of the descending willow roots into the well, a stone intervenes so as to obstruct the impingement of the watery vapor, the root divides into branches or turns to one side.

269. The Stem is that portion of the plant which

grows in an opposite direction from the root, and gives support to the leaves and organs of reproduction. All flowering plants have stems, though in some cases the stems are so short as not to protrude above the surface; such plants are called *stemless.* Stems do not always ascend; sometimes they trail on the ground, or burrow. At other times they are *creepers, climbers, twiners,* or *runners,* these terms indicating the particular disposition of the ascending axis. Again, when they are too weak to stand erect, they are said to be *decumbent,* and *procumbent* or *prostrate.*

270. *Creepers* run along the surface or just under the ground.

271. *Climbers* ascend by means of *tendrils* (the clasp of a vine), leaves or roots.

272. *Twiners* ascend spirally.

273. *Runners* shoot out leaves and roots at the extremities, as the strawberry.

274. *Decumbents* lean upon the ground, the base being erect.

275. *Procumbents* or *prostrates,* lie on the ground.

276. The *structure* of stems shows great differences in plants. It essentially consists of cellular and woody fibres embedded in cellular tissue, the whole covered by a skin or epidermis. But between plants which grow from seeds with one cotyledon, and such as grow from seeds with two cotyledons, there is a great difference as to the mode of organization and growth.

277. On account of their different modes of growth, stems are divided into two chief classes, namely, into *endogenous,* signifying to grow inwardly, and *exogenous,* signifying to grow outwardly.

278. The stems of monocotyledonous seeds, are endogenous. They present no distinct separation into pith, wood, and bark. The woody portion consists of bundles of fibres distributed throughout a cellular tissue, and not presenting any appearance of zones. Each of these fibres seems to vegetate separately, they are ranged around a central support, and are so disposed that the oldest are crowded outwardly by the development of new fibres in the centre of the stem. This pressure causes the external layers to be very close and compact. This mode of increase, little favorable to growth in diameter, produces long and straight stems nearly uniform in size throughout their whole extent, as the palms of the tropics, and the sugar cane and Indian corn of temperate latitudes. The epidermis is formed of the foot-stalks of leaves, which becoming indurated, remain upon the outer surface of the plant.

Fig. 22.

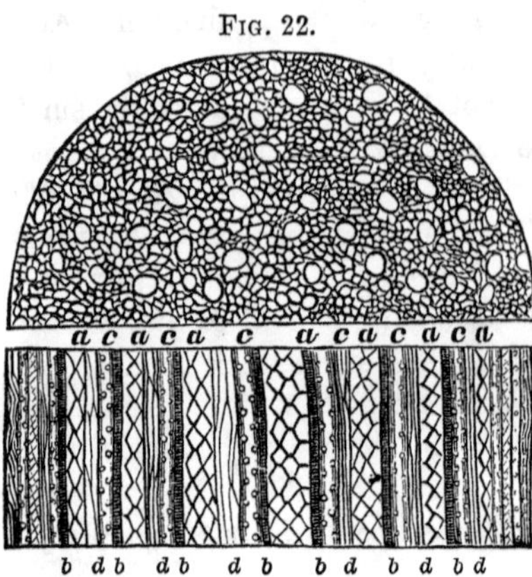

Section of Stem of Endogen.

a, cellular tissue ; *b*, spiral vessels ; *c*, dotted ducts ; *d*, woody fibre.

279. Dicotyledonous seeds produce *exogenous* stems. These are composed of three separate parts, arranged concentrically, namely, *bark*, *wood*, and *pith*.

280. The *bark* consists of an outer portion or cellular integument, composed of cellular tissue and covered by an epidermis, and an inner portion called *liber,* in contact with the wood. At certain periods, a mucilaginous product is interposed between the bark and the wood, which is supposed to be the material from which new cells are elaborated ; this layer is called *cambium.*

FIG. 23.

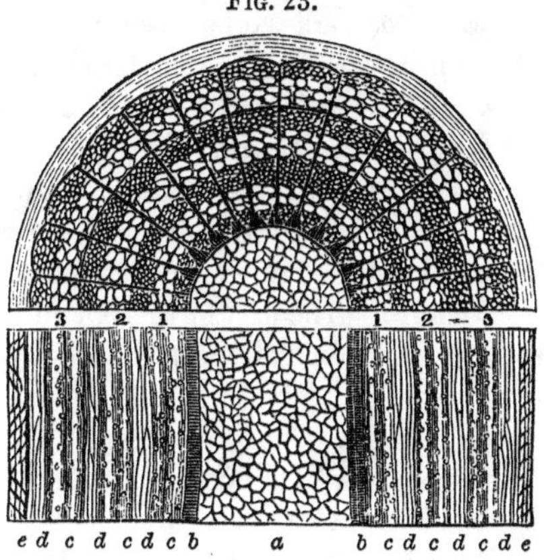

SECTIONS OF AN EXOGEN THREE YEARS OLD.

a, pith ; *b*, medullary sheath, exterior to which are three layers of wood—1, 2, 3—each formed of *c* dotted ducts and *d* woody fibre ; *e*, bark.

281. *Wood* consists of woody fibre, ducts, and cellular tissue arranged in zones. *Pith* consists of soft cellular tissue at first abounding with nutritive matter for the nourishment of the terminating buds, but afterwards becoming effete and dying. The pith is surrounded by a narrow zone of vascular tissue, which is termed the *medullary sheath ;* it consists of woody fibre and spiral vessels, and is the only part of the stem in which these latter occur. Both this sheath and the wood are traversed by narrow plates of condensed cellular tissue, passing from the pith to the bark, and denominated *medullary rays.* The number of zones or layers in a stem indicates its age, each layer being the product of a single year's growth. The woody portion is also

divided into two parts ; one exterior, new, colorless, and permeable to the circulating fluids, called, *alburnum* or *sap-wood ;* the other central, denser, charged with certain secretions, and impermeable,—this is termed *duramen* or *heart-wood.* Heart-wood is generally of darker color than the sap-wood, and as it does not assist in maintaining the functions of the tree, it may decay without any injury to the vitality of the plant. Each layer of wood consists of vessels or ducts and woody fibre. The first of these layers is the earliest formed, and, therefore, nearest the centre. Their relative arrangement, however, is not the same in different trees, and this difference is owing to the variation of cell development, which occasions the great variety of texture in different kinds of wood.

282. Buds (see Fig. 24) are nothing more than a repetition of the plant producing them, in an undeveloped state. The stem always commences from a bud, and terminates in one. Buds are commonly covered with rudimentary leaves, and those which vegetate in low temperature are protected by an additional covering of overlapping scales ; these scales, however, are rudimentary leaves. In high temperature where no such protection is necessary, buds are generally naked. We are accustomed to limit our appreciation of buds to branches and twigs, and not to regard them as independent individuals, which, in fact, they are, and not unlike the little nurslings of other beings, which so long as they remain in the paternal home continue in the closest connection with the parent. But their capability of independence may be shown in a variety of ways. The familiar operations of grafting and budding are examples of this, and the breaking off and sowing of the buds of forest

trees, when carefully attended, is a successful means of propagating plants which sometimes supersedes the more ordinary process. *Layering* is but another variety of the same thing. Both alike depend upon the facility with which these bud-plants develop adventitious *roots*, when brought in contact with the moist earth. But it is not alone by art that the multiplication of plants from buds is brought about. Some of our garden lilies produce little buds near the surface of the soil, which grow up to be independent shoots, bearing leaves, in the *axils* of which (the angles formed at the junction of the leaf with the stem), still stronger buds are produced, and these readily take root and produce vigorous plants. In this manner, too, strawberry buds soon cover a neglected garden. But of all buds, the *potato* is the best example of one capable of propagation ; this useful tuber being nothing more than a large subterranean fleshy bud. Other examples might be cited for showing the care taken by nature, that plants, especially those most useful for the food of animals, may be preserved under a variety of circumstances.

283. Leaves are membranous organs of various forms, growing from the stem or branches, and situated immediately below the buds. They consist of an extension of the skin or cuticle of the plant into a flat expanded surface, which is supported by a net-work of fibres and vessels derived from the medullary sheath. Leaves are said to be *alternate*, *opposite* or *verticillate*, according to position. And when they arise immediately from the stem, they are called *sessile*, when from fibres clustered together for a certain distance before they expand, they are *petiolate*, being supported by *petioles*—petiole being the name given to the footstalk

of a leaf. The leaf, therefore, consists of two parts, the petiole and the lamina or expansion. The projecting lines on the under surface of a leaf, are termed *veins*, and the distribution of these, upon which the form of the leaf depends, is called venation. In endogenous plants, the veins of leaves proceed directly from the base to the apex and are parallel with each other, hence they are denominated *parallel veined* leaves. In the exogenous plants, the veins are variously divided and form a net-work with extremities connected with each other in such a manner as to be appropriately named *recticulated* leaves. The internal structure of a leaf consists of two sets of veins, one belonging to the upper surface, and conveying the sap of the plant to the expansion; the other belonging to the under surface, and returning the elaborated juices to the bark. These two sets of vessels ramify through a cellular tissue, the parenchyma, which contains numerous green globules termed *chromule* or *chlorophyllin*, to which the color of the leaf is owing. In leaves as in buds, under some conditions, the propagation of species is provided for, independent of seeds. There have been instances where leaves lying on the ground, have become covered with buds, but the conjunction of all the favoring conditions for the development of a cell thus provided, is exceedingly rare. But there are some plants so organized as to make this a common means of propagation,—such is the case with the *bryophyllum calycinum.* In the leaves of this plant, embryo cells exist, and when placed upon the moist earth, these cells produce buds, and ultimately perfect plants. The leaves of the orange tree, also, frequently develop buds as do some other useful plants.

284. STIPULES are small leafy appendages at the base

of the petiole or sessile leaf, of the same structure as the leaf. And where their margins unite so as to form a sheath round the stem, they are called *ochreæ.* These are only to be found in endogenous plants.

285. Various modifications present themselves in the form of the leaf, and in the arrangement of the component parts ; but none of these affect the essential character of the organ as above described. But notwithstanding the various ways in which the leaves of plants are arranged on the stem, seem to present a great variety, they are all reducible to one fundamental type or plan, namely, a *spiral* which is the result of the *radiation* of the appendages, not from a single point, but from a longitudinal axis. So that any one flowering plant may be exhibited as the model of the whole series. To illustrate this SCHLEIDEN selected the *anagallis phœnicea,* or blue-flowered anagallis, as a type of the whole.

286. It will be perceived (Fig. 24) that the leaves come off at-regular intervals along the axis, but gradually out of line with one another,—the second not being above the first, but a little to one side of it ; the third holding the same relation to the second, and so on, in such a manner that a line carried through the points of origin of the successive leaves which are termed *nodes* will not only ascend the stem, but will gradually turn round it, and will at last pass through a point directly above the origin of the first leaf. Those leaves, which may be intersected by this line, whilst it makes one turn round the stem, are said to form a *cycle,* but the number of leaves which the cycle contains is subject to considerable variation. Dicotyledonous plants generally have *five* leaves to the cycle ; that is, the sixth leaf will be directly above the first, the eleventh above

the sixth, and so on. In monocotyledonous plants, the cycle commonly consists of *three* leaves; the fourth being above the first, the seventh above the fourth, and so on. The most common apparent departures from

FIG. 24.

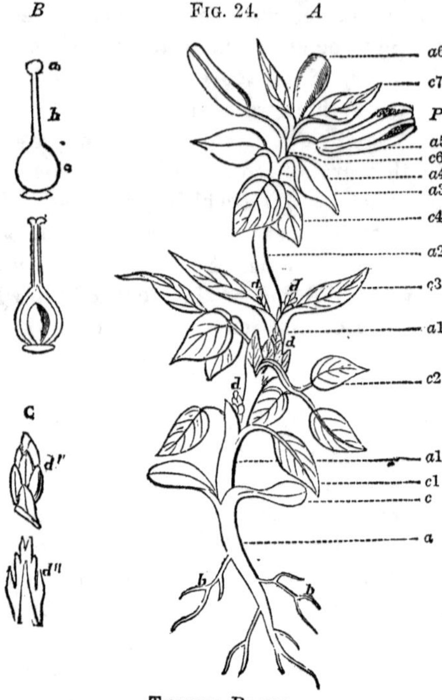

TYPICAL PLANT.

A, *a* to *a6*, the axis, *a* being the root; *a1*, *a2*, *a3*, *a4*, and *a5*, the successive internodes of the stem, and *a6* the terminal development of the axis into an ovule; *b*, rootlets; *c* to *c7* the successive foliaceous appendages to the axis, *c* being the cotyledons; *c1*, *c2*, and *c3*, the ordinary leaves; *c4*, the outer floral leaves or sepals, *c5* the inner floral leaves or petals; *c6* the stamens, and *c7* the carpellary leaves; *d*, leaf-buds. *P*, perianth. *B*, carpel inclosing an ovule, seen externally and in section,—*a*, the stigma; *b*, the style; *c*, the ovary. *C*, leaf-buds; *d'*, as seen externally, and *d''* in section.

the spiral type, are the opposite and verticillate arrangements. These are reconcilable, however, in various ways. The non-development of internodes (between the nodes), of the stem, for instance, being one way, which easily accounts for it. Suppose a cycle consists of two leaves ; by the non-development of every alternate segment or internode of the stem, the leaves would become opposite. In like manner the verticel of five leaves would occur by the omission or non-development of the internodes between five successive nodes.

287. The *perianth,* or floral apparatus, is composed of a series of verticels, which together with the intermediate portions of the stem, are comprehended in the term "*flower*" or "*blossom.*" In this may be distinguished, four degrees of development. The first, second, and fourth (*c*4, *a*5, and *c*7), differ only from the true leaves in the delicacy of their structure, and the second more especially by their color. These different whorls of the "flower," are called, *calyx, corolla,* and *carpels* or *fruit-leaves.* The fruit-leaves derive their name from the circumstance that in their subsequent changes they chiefly contribute to the formation of the fruit.

288. The third stage of development in the perianth (*P*), has totally different peculiarities, the leaf undergoes such important structural changes as to be scarcely recognizable. It becomes an elongated thick mass, containing several, usually four, longitudinal cases ; these become filled with a quantity of isolated dust-like cells, which by the opening of the cases are scattered in all directions ; these appendages are called *stamens ;* the cases in them *anthers,* and the cells in the cases, *pollen.*

289. In the place where the carpels or first-leaves are situated in the anagallis or *typical plant*, that is, in the centre of the flower, most plants display an organ which is closed in all round, but hollow ; it encloses the seed-buds, or germ-cells, and its cavity only communicates with the external air by a canal so small as to be almost imperceptible. This body as a whole is called the *pistil*, and in the foot of this, termed the *ovary*, is contained the *ovules* or germ-cells. The apex of the pistil is called the *stigma*, and the filiment connecting the stigma with the ovary, is termed style. The pollen grains of the stamen being scattered by bursting of the anthers, coming in contact with the stigma, emit special prolongations, called *pollen tubes*, and these convey the active *fovilla* or sperm-cells of the pollen to the germ-cells or ovules at the base of the style, contained in the ovary, by union of which the seed, or embryo-plant, is developed.

290. No part of the plant is so variable in structure as the style. Sometimes it is wholly formed of one or several carpels, at others, its ovary is a portion of transformed stem. Those portions of the stem, too, which otherwise belong to the blossom—the foot-stalk of the flower—are often strangely metamorphosed ; and on these differences depends in a great measure the variety of flowers, in this immense series of vegetation.

291. But the most variable manifestation of all, is that of the Fruit, a term applicable to all mature pistils, of whatever form, size or texture. Yet the accumulation of fleshy or pulpy matter in plants, commonly known as fruit, frequently takes place wholly independent of the pistil. The true fruit of the strawberry plant, for instance, consists of a number of

little hard granules never thought of as food, while the part so highly esteemed, is but a portion of the flower stalk. In a raspberry, on the other hand, we eat a quantity of small genuine fruits, while the flower-stalk, which is so delightful in the strawberry, is here the little white cone left on the stem. In the apple, again, we eat part of the flower-stalk ; in the cherry, part of a *leaf*, and in the walnut and almond, we devour an infant plant—root, stem, leaves, buds, and all !

CHAPTER XII.

MONSTROSITIES.

292. Anything out of the common order of nature, is occasionally designated by the term *monster.* And physiologists are wont to apply this term to animals in which one or more parts of the body present some congenital malformation. Thus limited in signification, scientific men once regarded "monsters" with feelings very little higher than those with which they are still looked upon by the ignorant; while the true physiologists now see in them the most interesting and suggestive illustrations of all that is beautiful in that unity of design which pervades all the purposes of nature.

293. Malformations are usually such, in virtue of their close conformity to the *general* model, and deficient only in those *special* developments which characterize a perfect species. Some of the most curious and interesting examples of this, are furnished by plants. There are families of plants distinguishnd no less by their *deficiencies* in conformity to the symmetrical arrangement portrayed in the last preceding chapter, than they are by apparently abnormal departures from any tendency to an ideal archetype. In the orders of plants known as *labiaceæ*, take sage, for example ;—the corolla instead of being a regular whorl, has two large

lips bounding its border; hence, they are denominated *labiate* corollas. And the stamens, instead of corresponding in number (five) with the sepals of the calyx, lack one, and the four are arranged into two sets,—two long and two short. In all which there is a departure from the *regular* flower. Again, in the snapdragon, *(scrophulariaceæ)*, a long spur is developed from one of the petals, whilst an upper lip is developed from another in such a manner as to form an arch like the mouth of the dragon! Other plants, instead of leaves. have *ascidia*, or pitchers, these are abnormal petioles.

294. But of all monstrosities in the vegetable world, the CACTUS FAMILY is the most remarkable; so "monstrous" are they, indeed, that one species (the cereus monstrosus), is named *deformed cactus*. The *cactaceæ* are all indigenous to America. But soon after the discovery of this country, their curious appearance at once attracted attention, and they were so rapidly distributed throughout the Old World, that they may now be considered as fully naturalized there. In America they are found between the parallels of 40° S. lat. and 40° N. lat., but in the greatest profusion on the sandy plains of the tropics.

295. Everything about these plants is wonderful. With the exception of a single genus, *peireskia*, no plant of the order possesses leaves Those parts of some species, the *alatus* and the *Indian fig*, which are commonly called leaves, are nothing but flattened expansions of the stem. On the other hand, they are all distinguished by an extraordinarily fleshy stem, which is clothed by a grayish-green, leathery cuticle. And in the places where leaves are situated in regular plants, the cactus is furnished with various tufts of hairs, spines, and prickles, in various degrees of

development. The *torch-thistles* rise in rounded or nine-fluted columns, to a height of forty feet, often without a single branch ; but frequently ramifying in the strangest ways, and sometimes elevating many armed branches resembling in their regularity an enormous candelabra. Other species there are, of much more humble aspect ; their globular, oval, or flattened branches piled up upon one another, or scattered at distances apart, are characteristically named *mammillaria, hedgehog,* and *melon.* Again, there are others with long whip-like stems, twisting snake-like among these, or climbing the adjacent trees, which have received the name of *serpent* cactus. The *torch-thistle* is a species of cactus, so called on account of one of the uses to which it is put, to light up the night. This is a small columnar species, which remains standing long after it ceases to live, and thus becomes dry and inflamable. This species is also used for building purposes, and seems specially adapted to the regions where it most abounds. The stems, by long standing, become so light as to be easily transported upon the backs of mules, to the heights of the Cordilleras, and there they are used for sills, beams, and posts, in the houses of the inhabitants, who would otherwise find it extremely difficult or wholly impossible to procure other timber.

296. The *Opuntia,* is especially remarkable for the long thick spines with which it is armed, on which account one species is appropriately named *Ferox,* the SAVAGE. This is commonly used in its native country to form fences ; and in the division of the Island of St. Kits by the English and French, a triple row of *opuntia ferox,* formed the boundary.

297. I have seen living hedges of the opuntia ferox,

in Venezuela, fifteen or twenty feet high; and each individual plant bearing spears interlaced with its fellows, so strong as to present an impassable defence. Within such enclosures are collected various flowering species of Cactus, together with other plants. Among these, the humble *Mammillaria* is adorned with flowers of beautiful purple and red colors; the *Speciossissimus* bears a large blossom of unequalled richness, resembling crimson velvet. And in the midst, the universally renowned CEREUS GRANDIFLORUS, or Night Blooming Cereus.

298. The blossoming of a Night Blooming Cereus, is always watched for with the most intense interest. When about to bloom, the bud begins to expand soon after the setting of the sun, and by the silent hour of midnight, the whole of its unequalled splendor is revealed. The calix is of a golden yellow color, and the petals of the purest white, the whole forming a gorgeous vanilla scented plume, more than a foot in diameter! But as if not to risk comparison with the matchless orb of day, ere the sun rises, the flower closes, never again to open.

FIG. 25.

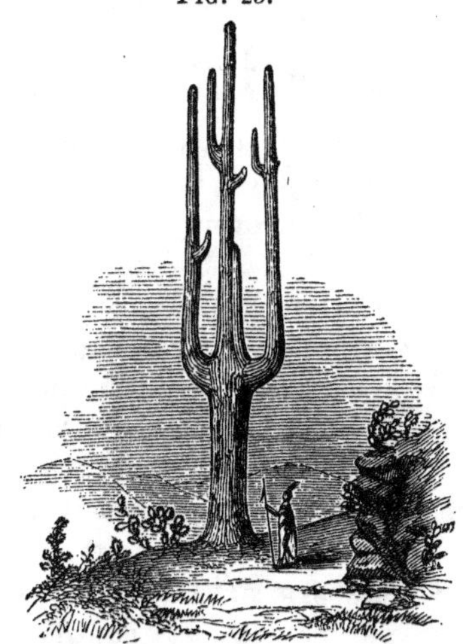

THE CEREUS GRANDIFLORA—Night Blooming Cereus.

299. The Cactaceæ are not alone interesting from the

strange contrasts they exhibit. Their economic uses are manifold, and besides those already referred to, nearly all bear edible fruits, and some of these are among the most delightful refreshments of the hot zones which ripen them.

300. The Opuntias, commonly called Indian Figs in the West Indies, furnish a delightful dessert fruit, and the little rose-red berries which are produced by the Mammillarias, in our hot-houses, have, in their tropical homes, a delicious acid flavor. Considered in relation with the climate where they attain perfection, the fruits of the *Cactaceæ* may be regarded as a nobler form of the Gooseberry and Currant of temperate latitudes, to which, also, in a botanical point of view, they are the nearest allies.

301. Like as many other plants, the Cactaceæ support an *animal* of great importance. This is the *Coccus cacti*, or Cochineal.—A little more than a century ago, there was much controversy in Europe, whether the Cochineal was the seed of a plant, or an insect. The Spaniards who first settled in Mexico, and subsequently the Mexican government, having kept the nature of Cochineal secret, confined its cultivation exclusively to Mexico until 1785. In that year, De Mononville, at the risk of his life, succeeded in carrying some of the live insects to the French colony of St. Domingo. Since then, it has been cultivated in the Canaries, Corsica, Spain, and other places. But Mexico still produces the finest kind, and the greatest quantity.

CHAPTER XIII.

TRANSITION OF MATTER.

302. The life of Plants is closely connected with the life of Animals in a most simple manner, and for a wise and sublime purpose. In explaining natural subjects and causes, the simplest incidents, assisted by the science of quantity and number as auxiliaries in the study of the more complex principles as variously exhibited, are not only all that is necessary, but far more available in eliciting as well as in applying the truths we would inculcate. And in proportion as we pursue Nature's simplest truths, we perceive the utility, and then the necessity of her arrangement, in adapting means to ends, which, on first glance, and without due consideration, presents nothing but disorder and confusion.

303. A different arrangement only of the elements of organic bodies, produces substances totally different in both structure and properties. In the processes of death and decomposition, all animal bodies are dissolved into ammonia, carbonic acid, and water; and these exhale as gas and watery vapor into the air, whence the plants take possession of them, and form compounds suited to the sustenance of animals. Yet these matters can never so combine, as not to contain far more oxyen than is essential for plants. Free

oxyen gas is thus continually given off, and the atmosphere is maintained in a state of purity alike adapted to the well-being of both plants and animals.

304. Carbonic acid, ammonia, and watery vapor, are essential food for plants, and on these elements vegetation at first lived. But when, in the order of Providence, Man was necessitated to till the earth, we find plants possessed of qualities which do not belong to them in their wild state, and these qualities are exactly those which render them fit for food. The branched stem of the wild Cauliflower, is a tough, bitter herb, wholly different from the white, tender aromatic plant, known by the same name when subjected to cultivation. And the wild Carrot is a dry slender root, wholly unfit for food, while, under cultivation, it becomes large, juicy, and nutritive.

305. All the various properties through which certain plants have become of so much importance to Man, have been called forth by circumstances originally foreign to the Plant. The necessary conditions lie not in the elements which are the same for all plants, but in the organic constituents present in the soil, and taken up by the roots. Wherever the soil is rich in the various salts occurring most abundantly in plants, the characters of these latter become altered ; varieties and monstrosities originate, which never occur in the wild condition, where the plant always keeps to the soil exactly agreeing to it.

306. Plants, however, exhibit varied dispositions to the alteration of their peculiar nature, by the influence of cultivation. While some retain the most minute characters under diverse conditions, others run into innumerable varieties. In others, the varieties exhibit but little stability, quickly returning to the wild form,

or new variations. Others, again, produce manifold aberrant forms, which after some years' culture, may be propagated with full certainty by their seeds, and thus give rise to what are called sub-species. It is this peculiarity of plants, which fits them to become advantageous objects of cultivation; and it is in virtue of this quality, that they readily produce very different and stable varieties, out of which Man selects those most profitable for his purposes, and receives them into the number of his vegetable subjects.

307. All such parts of vegetables as can afford nutriment to animals, contain certain constituents which are rich in nitrogen, and experience proves that animals require for their support and nutrition less food, in proportion as it abounds in nitrogenized constituents. These constituents are especially abundant in the seeds of the different kinds of grain, and of peas, beans, and lentils; and in the roots, and the juices of what are commonly called vegetables. But they exist in all plants without exception, and in every part of plants in greater or smaller quantity.

308. The nitrogenous compounds in the vegetable kingdom may be reduced to three substances, namely, *vegetable fibrin*, *vegetable albumen*, and *vegetable casein*, which are easily distinguished by their external characters. Two of them are soluble in water, the third is insoluble.

309. When the newly expressed juices of vegetables are allowed to stand, a separation takes place in a few minutes. A gelatinous precipitate, commonly of a greenish tinge, is deposited, and this, when acted on by liquids, which remove the coloring matter, leaves a grayish white substance. This is *vegetable fibrin.* The juice of grapes is especially rich in fibrin, but fibrin

is most abundant in the seeds of wheat and other cereal grains. The method by which it is obtained shows that it is soluble in water, though originally in a state of solution in the vegetable juices.

310. The second nitrogenized compound remains dissolved in the juice after the separation of the fibrin, or it may be obtained from the clarified juice of nutritious vegetables, such as cauliflower, asparagus, turnips, etc. It does not separate from the juice at the ordinary temperature, but when the liquid containing it is heated to the boiling point, a coagulum is formed, which in all respects resembles that produced by heating to the boiling point the white of egg diluted with water. This is *vegetable albumen.* It is found most abundantly in certain seeds—nuts, almonds, etc.

311. The third nitrogenized constituent of vegetables is *vegetable casein.* It is chiefly found in leguminous seeds—peas, beans, lentils, etc.; like vegetable albumen, it is soluble in water, but differs from it in this, that its solution is not coagulated by heat.

312. These three compounds, vegetable *fibrin, albumen,* and *casein,* are the true nitrogenized constituents of animal food; all other nitrogenized compounds occurring in plants, are either rejected by animals as poisonous, or else they occur in the food in such small quantities as to be of no utility in the nutrition of the animal body. These three substances all contain the same organic elements, united in the same proportion by weight, and what is still more remarkable, they are identical in composition with the chief constituents of blood. These vegetable principles, which in animals are used to form blood, contain the chief constituents of blood already formed! It follows, therefore, that the development of the animal organism and its growth

are dependent on the reception of certain principles from the vegetable, identical with the chief constituents of the blood, already formed. The animal must, indeed, receive these substances *ready prepared,* in order to apply them to its nutrition.

313. Albumen, fibrin, and casein, are, therefore, the exclusive *materials for nutrition;* they cannot be replaced by any other substance, and if they are entirely withheld, the body must necessarily die of starvation.

314. Besides these nitrogenized compounds necessary for the *nutrition* of the animal body, other compounds *devoid* of nitrogen must also be present, as it were, for fuel on the hearth of organic life. These are denominated *materials of respiration.* Among these, the chief are gum, sugar, starch, and the liquors prepared from them—beer, wine, distilled spirits—and fat. These substances merely pass through the animal body, the carbon and hydrogen being taken up by the oxygen received in respiration, are expired as carbonic acid and water.

315. Comparing the requisitions which the animal body makes in behalf of its maintenance, with the components of plants which serve for food, we find in all plants a certain amount of albumen dissolved in the juices. In the cereal grains and buckwheat especially, the substance commonly called glutin resembles a mixture of gelatin and animal fibrin. The leguminous seeds contain a substance resembling animal casein. The class of plants furnishing the materials of respiration, are no less widely distributed throughout the vegetable world.

316. When we review the nutritive substances which man obtains from the vegetable kingdom, we find them

conveniently classible into three groups. The first group is remarkable for the great quantity of starch contained in the plants composing it. To this group belong the Cereals and Pulses, the tuberous vegetables,—Potatoes, Yams, Mandioc, Sago Palms, etc. The second group includes the fruits, rich in sugar and gum, which owe their peculiar cooling properties to malic, citric, and tartaric acids, and their delicious flavors to the presence of aromatic substances ; of such are our well known fruits. And, in addition to these, may be named the Banana, Date, Bread-Fruit, and the juicy stem of the Sugar-Cane ; also, the saccharine and gummy roots, which constitute a large proportion of our kitchen vegetables. The third group, consists of the oleaginous kernels of various fruits,—the Cocoa-nut, Brazil-nut, Almond, Pecan-nut, Walnut, and many others. To the foregoing, may be added all the various beverages, which are nearly all derived from the vegetable kingdom.

317. In the endless series of compounds which begins with carbonic acid, ammonia, and water, the sources of the nutrition of vegetables, including the most complex constituents of the animal body, there is no blank, no interruption. As soon as the Plant has borne seed it dies, or, at least, a period of its life comes to a termination ; and the first substance capable of affording nutriment to animals, is the last product of the creative energy of vegetables. The development of the Animal begins on those substances, with the production of which the life of the Plant ordinarily ends.

318. We have seen that the nutrient fluids contained in the animal body, and the gelatin which forms the basis of bones, are essentially produced from substances containing nitrogen, furnished by plants. But

gelatin alone is insufficient for the bony structures,—these contain, besides gelatin, bone-earth, a compound of carbonate and phosphate of lime. It is to this substance, that bone owes its hardness and durability, through which it is fitted to become the foundation and support of the animal body. A deficiency of bone-earth in the human system is followed by a dreadful disease, the rickets. The fluids of the animal body likewise contain *salts,* without which they cannot execute the functions to which they are appointed. Whence, it may be asked, do these things come, if the whole Animal World is nourished on the Vegetable Kingdom? Above all, whence come these earths and salts, if the Vegetable Kingdom is nourished only by Oxygen, Carbon, Hydrogen, and Nitrogen? The loathsome habits of the dirt-eating Ottomacs, and of the clay-eating Negroes, force themselves upon us in this relation. But these are abnormal phenomena, proceeding from diseased conditions, or starvation. And we are again compelled to return to the PLANTS for an answer.

319. Strange to say, among the *materials for nutrition* in plants, we find *phosphate of lime* constantly associated; and among the *materials for respiration,* devoid of nitrogen, we find *alkalies.* Nor is this all; we find these substances in plants in exactly the same determinate combination as are suited to the constitution of animals. Supported by these facts, Liebig has shown that, since organic nutriment stands everywhere in equal abundance at the service of all plants, the cause of the great difference of vegetation cannot be sought in the nature of the plant, but in the *inorganic* elements of the soil, and that it is essentially indifferent whether we convey manure to the field, or burn it first, and strew the ashes on the soil, since its efficacy is solely

dependent upon the constituents of the ashes. From these premises, we can easily deduce most profitable results. We can conceive why an *irrigated* rice-field in South Carolina can produce a large annual yield without manure, since the necessary quantity of salts are brought to it by the spring water. It becomes clear how the luxuriant harvest of Indian corn is reared on the arid sand-drifts of Peru, watered by the little rills, fruitful in soluble salts from the snowy peaks of the Andes. Hundreds of similar phenomena overwhelm the admirer of the handy-work of Nature in wonder, love, and praise!

320. The ashes of plants are characterized by four chief classes of constituents: readily soluble alkaline salts; earths, especially lime and magnesia; phosphoric acid and silex. Sometimes one, sometimes two of these substances predominate in the ashes of the plant. According to this, Liebig divides the cultivated vegetables into:

I. Alkali plants, to which belong Potatoes and Beets.
II. Lime plants—Clover, Peas, etc.
III. Silex plants—the Grasses.
IV. Phosphorus plants—conprehending Rye and Wheat.

And besides these, plants contain many other substances, the importance of which we do not yet fully understand. In the progress of agricultural science, however, new properties of the vegetable world are constantly opening new rewards to the faithful follower of the resources of nature.

321. Oxygen, carbon, hydrogen, and nitrogen in plants, cannot form from them a single granule of albumen or gluten, unless they are associated with the salts of phosphoric acid. The useful starch, the sweet sugar, the cooling citric acid, the aromatic oil of

oranges, are solely composed of carbon, hydrogen, and oxygen, but the plant cannot prepare those gifts for us out of ever so great an abundance of these, if it does not possess also alkaline salts. The slender stalk of the wheat could not lift its fruitful head to meet the sun's rays, unless the soil furnished it with silex, through which its cells obtained the necessary solidity to maintain an erect position.

322. In view of the foregoing premises, it is the peculiar province of the scientific agriculturist, to furnish to every kind of soil and plant, a proper compost of those mineral substances which the plant requires and the soil is deficient in, and in such a state of combination that the substances shall be soluble enough to be taken up by the plants, and yet not so readily soluble, that the rain can wash away any considerable quantity. But the inorganic compounds are inseparably connected with the organic in the plant, and if we would take possession of the latter, we must also have the former; we must constantly bear in mind that without *humus*—though of itself innutritious—there can be no healthy and strong vegetation.

323. It is not alone sufficient that the different substances necessary for healthy vegetation be abundant in the soil,—they must be present *in the proper proportion* to each other. This condition is of the highest importance, in reference to those plants which are naturally inclined to produce varieties; above all, in reference to those plants, the chemical composition of which renders them most liable to essential injury by alteration of their constituents. All this especially concerns the potato, but does not much affect wheat, rye, and other grains. When we compare the constituents of the ashes of grain with the contents of freshly

manured soil, we find the proportionate constituents of the two very nearly alike, and what is very remarkable, when the constituents of the ash of rye are abstracted from the soil, the matters left behind are almost identical with the ash of the potato. It may, therefore, be set down as an important practical deduction, that the *potato should never be cultivated as the first crop*, but should succeed the cultivation of rye, wheat, or clover. And that where the potato crop fails, the soil should be appropriated to other produce, for such time as may be necessary to refit it for the *healthy* potato.

324. The wild potato is a small, greenish, and bitter-flavored tuber, containing a great deal of starch. It is one of those plants which readily produces varieties under cultivation, which exhibit a good deal of permanence when the conditions of culture remain exactly the same; when this is not the case, new varieties arise —they "sport," as it is said. The difference of these varieties consists in part in the alteration of the form of the potato, and in its quicker or slower maturity. But much more important is the difference in the chemical process by which the relative amounts of starch and albumen become altered. Starch, a substance containing no nitrogen, is the peculiar characteristic constituent of the potato, a substance which withstands decomposition for a long time. The formation of this requires the presence of a large quantity of potash, and hence it is, that the potato belongs especially to the alkali plants. The albumen of wheat, on the contrary, is rich in nitrogen, and is particularly prone to decomposition; its presence in large quantity renders the other substances—cellulose and starch—which can alone long resist decay, also liable to decomposition.

The production of albuminous plants pre-supposes the presence of a great quantity of the salts of phosphoric acid.

325. In the healthy normal potato, the proportion of the nitrogenous to the non-nitrogenous constituents averages 1.20 ; the proportion of phosphoric acid salts is as 1.10. Whereas freshly manured cultivated land, contains the inorganic constituents mentioned almost in the proportions of 1.2. The consequence of this is, that in such soils the plant is forced to take up the phosphates in larger quantities, in proportion to the alkaline salts, than its nature requires, and thence a greater abundance of nitrogenous matter,—of albumen,—is formed in it, than it would contain in a normal condition. This nitrogen renders the components of the potato, which always contains a great deal of water favoring it, prone to decomposition. This may appear under varied forms, as in the dry-rot, formerly observed, principally seizing upon the starch ; or as in the moist-rot, especially attacking the cellulose. That such a disposition may show itself as a ruinous disease, particularly in unfavorable weather, is readily conceivable, and it stands to reason that when the injurious influence which produced the seeds of the disease continue, the degeneration of the potato, and its proneness to disease, must be always increasing.

326. But in the superior claims of our own domesticated vegetables, we must not lose sight of the great food-plants of the desert, and of the world;—those which are prepared by nature for the sustenance of animals to the end of time. The history of our chief food-plants covers but a comparatively short period, and it is surprising to reflect upon the circumstance, that but a few species of a single family of plants furnish the

principal food of the greater proportion of civilized nations. The corn plants, or *cerealia*, of the family of grasses, includes about 4,000 species, and yet not twenty of them are cultivated for the food of man! The other food-plants, commonly used in temperate latitudes consist of a small number of subterraneous tubers, which send shoots above the soil, persisting but a few months, when like the cereals they cast their seed and die.

327. In contrast with the foregoing, the Bread-Fruit, Cocoa-nut, and the Date, furnish the chief food to great bodies of men over widely extended areas. To these may be added those plants which yield milky-juice. The chief of these used for food, are the *Tapioca Mandioc*, and Cow-tree or *Palo de Vacca*, of South America. When a tolerably large incision is made in the trunk of this tree, a white, oily, fragrant, and sweet fluid, similar to cow's milk, flows out in sufficient quantity to refresh and satisfy the hunger of several persons. To the same series belong all those useful plants which yield *Caoutchouc*, now manufactured into articles of universal utility But of all food-plants, the BANANA is the first and most valuable gift of Nature. Its slightly aromatic, sweet and nutritive fruits, are the principal food of the greater part of the inhabitants of all tropical latitudes. A creeping, subterraneous root-stalk sends up shoots twenty feet high, which consist merely of rolled-up sheath-like leaf-stalks, bearing leaves ten feet long, and two feet broad; the midrib of the leaf is firm and thick, but the blade on either side of it is so delicate, that it is readily torn by the wind, whence the leaf acquires a feathered aspect. In the midst of a thick bundle of such leaves, rises the rich clustered flower-stalk, which, within three months, bears from one hundred and fifty to two hundred mature fruits,

that in the aggregate weigh from fifty to seventy-five pounds. A space of ground which in six months would produce one thousand pounds of Potatoes, in half that period of time produces forty-four thousand pounds of Bananas. And if we take account of the nutritious matter which this fruit contains, a surface which, sown with wheat, would feed one man, planted with Bananas, affords sustenance to twenty-five men !

328. Nearly allied to these prolific vegetables in food for Man, are the giant Baobab, or Monkey's-bread, of a thousand years growth for other animals ; and the Cactus plants, which have been not inaptly termed by St. Pierre, the "Springs of the Desert." By the succulency of their tissue, they provide themselves in the rainy season with fluid sufficient to brave the long droughts of a tropical sun. This peculiarity gives them inestimable value to the fainting traveler. The wild Ass of the Llanos, too, knows well how to avail himself of these plants. In the dry season, when all animal life flies from the parched Pampas, when Cayman and Boa sink into death-like sleep in the dried-up mud of river beds, the wild Ass alone traversing the steppe, seeks the Cactus to quench his thirst ; cautiously stripping off the dangerous spines of the *Mellocactus* with his hoof, he sucks in safety the cooling juice !

329. The Cactus plants serve to explain what was for a long time a disputed point in vegetable physiology, namely, the manner in which plants obtain the nutritive fluids which pervade their tissues. The conveyance of nutrient fluid to remote parts of the organism of plants, varies somewhat according to the grade of organic development. In the lowest grades of the Cryptogamia, where there is no tendency to elongation in any particular direction; fluid is absorbed and trans-

mitted equally on all sides. But where plants are separated into different parts or organs, absorption is restricted to a particular portion of the surface. The evolution of an axis formed of prolonged cells, shapes the direction which the fluid takes to every part of the system,—the movement being most rapid where there is least resistance, just as fluid penetrates paper of different degrees of porosity. This process applies to the lower Cryptogamia, some of which are soft and succulent, others hard and crispy.

330. In the Phanerogamia, the fluid imbibed by the roots is carried into the stem, where its receives the name of SAP. Each annual layer of the stem of exogenous plants, consists of woody fibre and ducts intermixed. Through these ducts or interstices the sap ascends. In endogenous plants, the sap-tubes are mixed through the whole stem.

331. *Why* the fluid absorbed by the roots of plants ascends, is to many a puzzling question. It is owing to two causes, which may be illustrated in this wise. If the stem of a young plant be cut off in the spring, and the divided extremity immersed in water, it will absorb a sufficient quantity of fluid for the temporary support of the leaves ; whilst, on the other hand, the divided stock left standing will pour out fluid from the top forced up by the roots. The explanation of these two forces is, that in the first place the succulent extremities of the roots serve as the medium of the whole process, and the quantity of fluid absorbed by the roots, is always in proportion to the rapidity of its removal by the leaves. But the propulsive power of the roots exhausts itself in raising the sap to the sphere of the leaves, after which the leaves continue the force by the power of chemical attraction,—just as the con-

tinued rise of oil in the lampwick is regulated by the rate of combustion at its apex. On ascending to the leaves, the nutritive matters of the fluid are ebaborated or separated, and the remainder is then drawn off by exposure to the air. For this remainder, the plant no longer has use, and it is therefore driven off by the superior attraction possessed, in the first place, by the roots, and, secondly, by the whole plant,—for another portion of fluid is ready to undergo the same changes, to be in turn rejected for a new supply. The nutritive matters separated by the leaves from the liquid absorbed by the roots, constitute the *proper juice* of the plant. This is taken up by the vessels on the under surface of the leaves, and distributed to every part of the fabric, and it is by this fluid that the plant is nourished, and yields nourishment to animals.

332. The assumption that plants imbibe their sole nourishment from the air, is only consistent with the structure of the protophytes and other plants, which first grew upon the earth. That other plants are equally capable of this mode of nourishment is no less inconsistent with their structure than with their function in the organic world, for it is impossible that the inorganic elements which combine in their tissues, should exist in a state of vapor in the atmosphere. But it is believed by no less an authority—*in chemistry*—than Liebig, that from the vast amount of watery juice in the cactus tribe,—joined to the fact that most of them, and those richest in sap, vegetate on dry sand almost devoid of humus, and where they are, besides, exposed often three-fourths of the year to the parching beams of a tropical sun—draw their nourishment from the air. In proof of this position, the circumstance has been cited that the separated branches of the cactus stems frequently grow and produce shoots. By

weighing such shoots which have grown without soil, they will always be found to have become lighter, conclusively proving that instead of abstracting anything from the atmosphere, they have rather yielded something to it. All the growth that takes place in such cases, is at the expense of nutritive matters previously accumulated in the tissues.

333. The stem of the cactus being devoid of leaves and clothed with a dense leathery bark, which performs the double purpose of both leaves and bark, is peculiarly unfavorable to the evaporation of sap, and it is doubtless due to this peculiarity of structure, that they so long retain the fluid absorbed at a propitious season.

334. In order to determine the laws which govern the changes that take place in organic bodies, it is necessary to collect and compare all the facts which we can draw from an extended observation of the phenomena of life, in various orders of organic structure. The changes which occur in the life of any one being, are inadequate to furnish the required information to understand all beings, since any one body only presents us with a group of dissimilar phenomena, incapable of comparison with each other. Were we to base our notions of vegetable physiology on the history of any one plant, we should obtain but vague ideas of its different nutritive processes, since we cannot separate these processes one from another, so as to study them apart. We should be equally apt to form erroneous impressions of these same processes in relation to the animal body, were we to undertake to study them as they exist in the human system, without first passing in review the functions of those similar beings of the animal kingdom, which serve to illustrate the perfection of development as it exists in the erect stature and expressive countenance of man.

CHAPTER XIV.

THE PROTOZOA.

334* On the 26th of January, 1843, a crowd collected at the Round Down Cliff, near Dover, in anxious expectation, to witness the effect of the grandest and most daring display of blasting ever attempted by the skillful combination of human ingenuity. The labor of years had been expended on the preparations, in the openings of shafts and galleries for the immense undertaking, which in a moment of time blew up the enormous cliff and hurled it into the sea! In less than one minute, millions of tons of chalk were torn away, and a surface of fifteen acres covered with the fragments. From this may be estimated the tremendous power which must have been exerted. And with what did the combined ingenuity of the human mind enter into this giant struggle? With the remains of creatures, a thousand of which might be crushed by the pressure of a finger.

335. The ocean grove is without verdure, yet there is full compensation in its perpetual bloom. Each coral branch is every where covered with its star-shaped animals,—the coral blossoms. How the "tree of stone" grows and spreads its branches, what its connection with the blossoms on its surface, and whence the lime that constitutes the Island Cliff, or the

Fig. 27.

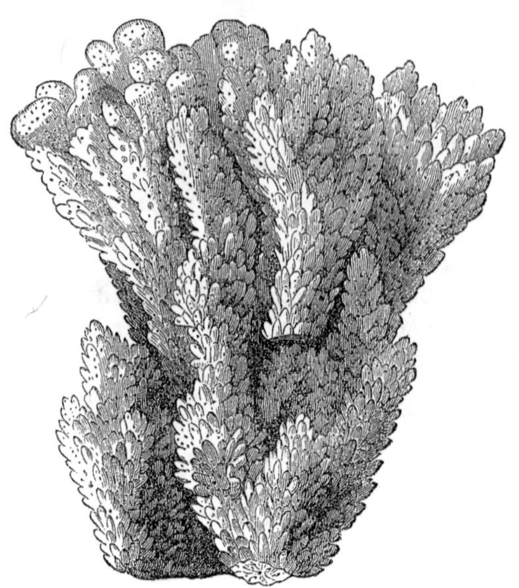

Coral.

rock of marble, which the human artist carves to delight intellectual faculties, and how this artist is formed, are questions which can only be answered by an acquaintance with the cell that enters alike into the constitution of the lowest and the highest forms of organization.

336. The history and appearance of the animal-cell in its simplest form, is essentially that of the vegetable of the lowest kind,—every one living for and by itself. It was an axiom of the celebrated Linæus, that "Stones *grow*, Vegetables *grow* and *live*, Animals, *grow*, *live*, and *feel*." Capability of feeling, therefore, was the final

characteristic chosen by Linæus, whereby an animal could be distinguished. But there are animals totally devoid of sensation ; they may be torn, or cut, or pierced with a red-hot iron ; or they may be acted upon by chemical stimuli of every kind, without the slightest indication of *feeling*. While, on the other hand, we have the sensitive plant, which shrinks from the slightest touch. If, therefore, we were to estimate the sense of feeling as manifested by motion on contact, there are plants which possess a far higher claim to the title of animals, than many growing and living creatures, which are wholly incapable of feeling. However manifest the differences between the higher forms of vegetable and animal beings, they are in their simplest forms incapable of being distinguished the one from the other. And the utmost of our investigations in this particular, only establish the truth that every plant and animal is formed after a general plan, while it is intended all along by its MAKER for a special end and no other, and it is only *as it advances* that we can discover that end.

337 If, for illustration, we enter a ship-yard, we discover a community of place in the materials there gathered together for use, and in the laying of the keel of every vessel. But it is only as the fabric advances that we begin to appreciate what was known from the first by the builder—the special purpose for which the ship is intended ; whether it is to be propelled by sails or by steam ; whether it is meant for warlike or peaceful purposes ; whether it is to convey articles of commerce or passengers. We only knew at the *beginning* that it was a *vessel of some kind*.

338. "Light and darkness are distinct from each other, and no one possessed of eye-sight would be in danger

of confounding night with day, yet he who, looking upon the evening sky, would attempt to point out precisely the hue of separation between the parting day and the approaching night, would have a difficult task to perform. Thus it is with the physiologist who endeavors to draw the boundary between these two grand kingdoms of nature, for so gradually and imperceptibly do their confines blend, that it is at present utterly out of his power to define exactly where vegetable existence ceases and animal life begins."* It is, indeed, quite evident that there are no certain means of distinguishing the lowest orders of plants and animals from each other, and we are at last reduced to the necessity of adopting that division of the kingdom of nature with which we set out, namely, inorganic and organic bodies only; and if we would subdivide this latter class, and allot one part to the botanist and the other to the zoologist, our chief concern is to know at what degree of organic development the distinction can certainly be made.

339. The first difference relied upon by modern naturalists, by means of which animal bodies may be distinguished from vegetable, is in the appearance of the *tissues* or component parts under the microscope. And when the process of development is such as to enable us to apply this means of distinction, we discover the simplest forms of animal bodies to consist of single cells, or of aggregations of cells, all of which are alike; and in the lowest forms of them, there is not even that distinctness of the cell-wall from the cell contents which exists in every perfectly formed cell, but the whole forms one *mass of living jelly* Animals

---

* T. Rymer Jones' *Natural History of Animals.*

thus constituted, are precisely parallel to the *protophyta* of the vegetable kingdom, and these are, therefore, appropriately designated PROTOZOA or First Animals.

340. SPONGES are the lowest known order of organization to which the name of animal is applicable. The substances known in commerce under the name of sponge, were formerly deemed of vegetable origin But modern naturalists assign them to the sphere of the zoologist. And since the investigations of the distinguished English professor of natural history, Dr. Grant, they are universally classed as animals. They consist of a film of gelatinous substance without feeling, possessing no stomach, and spread out upon a framework of their own construction ; they have the power of nourishing themselves and of separating from the sea, in which they are immersed, particles of a horny, calcareous, or silicious nature, and of building up by means of these materials a peculiar structure, called SPONGE.

FIG. 28.

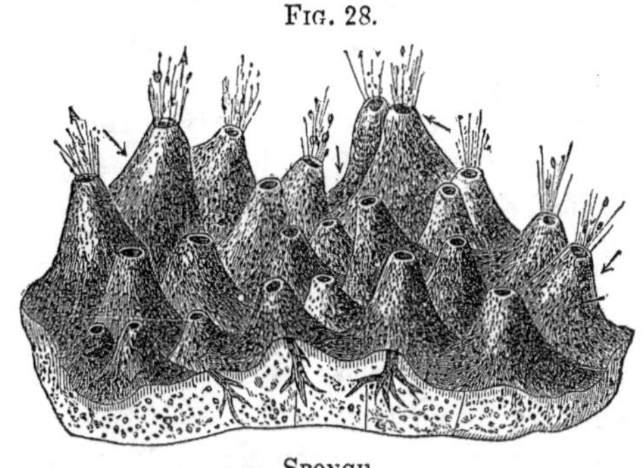

SPONGE.

341. By a careful examination of living sponges, Dr. Grant ascertained that the water wherein the sponge is immersed is perpetually sucked into its substance

through the countless minute pores that cover its outer surface, and is as incessantly expelled through other and much larger orifices. Having placed a portion of living sponge in a watch-glass with some sea-water, "I beheld," says he, "for the first time the splendid spectacle of this living fountain vomiting forth from a circular cavity an impetuous torrent of liquid matter, and hurling along in rapid succession opaque masses which it strewed everywhere around. The beauty and novelty of such a scene in the animal kingdom long arrested my attention; but, after twenty-five minutes of constant observation, I was obliged to withdraw my eye, from fatigue, without having seen the torrent for one instant change its direction or diminish the rapidity of its course."* It is by means of this incessant circulation that particles of organized matter suspended in the water are applied to the nutrition of the sponge, and the intricate framework or skeleton commonly known by that name built up. On close examination of the opaque masses, which constantly escape with the issuing torrent, they are found to consist of minute masses of jelly. These are, on close inspection of the living skeleton, found sprouting on its surface. They gradually increase in size, until about as large as pin-heads, when they become detached from the nidus where they grow, and are drawn into the currents by which they are whirled away and projected from the *parents*, for they are, indeed, so many young sponges.

342. The adult sponge is cemented to the rock, fixed and immovable, and consequently incapable of carrying the offspring from place to place, or in any manner

* Jameson's *Edinburgh Philosophical Journal*, Vol. XIII.

aiding them in search of a resting place. In lieu of this the offspring is endowed with organs of motion. Before being detached from the parent, they become pyreform, and while the smaller extremity (*b*) remains attached, the larger develops an immense number of microscopic filaments, resembling hairs, and named *cilia* (*a*); these are observed to be in constant vibration. And when the young sponge or gemmule is cast off, it rows away, as it were, with thousands of almost invisible oars. Having at length attained sufficient magnitude to take on the proper functions of the parent, it finally settles upon some congenial rock, and there agglutinates for the remaining period of its existence.

FIG. 29.

a

b

YOUNG SPONGE.

343. As the sponge increases in size the space for imbibition is proportionately enlarged, and the centrifugal water, flowing at first into one current, afterwards divides into many currents, and these gradually form channels in the cellular texture, the fibres of which are kept apart by the continuance of the stream. The channels being thus multiplied by the increased magnitude of the mass, at last begin to open into and cross one another, and so establish an intercommunication with each other throughout the tissue. But the issuing currents continually extends the superficies, and, as the ducts are drawn together near the upper surface of the sponge, many of them unite and form a common outlet, and so produce the single opening commonly existing at the apex. The shape and size of this opening, or *osculum*, depends in a great measure upon the texture of the particular species of sponge, it being larger or smaller, coned or flat, according to the compactness of the fibres.

344. The sponge of commerce, consists of a flexible horny substance, of such peculiar resiliency, as to be exceedingly useful for a variety of purposes. But there are other kinds, utterly worthless in a commercial point of view; the skeletons of these being formed by calcareous or silicious matter, deposited in a crystallized form throughout the entire structure, and imbedded in a tough fibrous material that binds them together in a hard unyielding mass. Unimportant, however, and worthless as such sponges are in the arts, they serve to throw LIGHT ON LIFE.

345. At the magnificent blasting of Dover Cliffs great masses of flint were discovered imbedded in the mountain of chalk. Flint, also, exists in greater or less quantity in most chalky formations, and these flints were once sponge! Sponges of this character, like those more highly prized, being separated as little gelatinous gemmules, attach themselves to the appropriate fastening places prepared by their cotemporary *confrères*, the coral polyps; and long after they have ceased to live, according to a well-known law of chemistry, particles of similar matter continue to be aggregated. And thus the little silicious spicula that originally constituted the framework of a sponge, have formed nuclei, around which kindred atoms continue to accumulate. Imbedded in the substance of the chalk, which, during long periods, successively overwhelms many generations of marine animals, the sponges remain for centuries exposed to the water that incessantly percolates such strata; the water all the while adding new particles of silicious matter, which it holds in solution. In the lapse of ages, the silicious spicula that originally constituted the framework of a sponge, have aggregated masses of kindred matter, which, at

the last, is converted into solid flint. It is common to find in chalky districts, flints, which, on being broken, still contain portions of the original sponges in an almost unaltered condition, and thus afford proof of their original state.

346. There are many other animals of the same grade, in the same degree of development as the sponge, consisting of mere masses of jelly, of uniform texture throughout, without a stomach, and incapable of feeling. Like the sponge, they may be torn, cut, or pierced with red-hot iron, without the least manifestation of any other function than that of contractility, which is common to many Plants. Yet these simple bodies fill an order in Nature, and the utility of the simplest animal known, the Sponge, finds its counterpart in any of the various other parallel forms of organic nature. Each one fills its sphere.

347. The Actinæ.—Animals of the lowest degree of organization are commonly classified by Naturalists as the Actinæ, from *aktin*, ray of the sun, because they are usually supplied with *rays*. Some, however, are destitute of rays, and on that account the term Acrita, from *akrino*, not to be discerned, has been suggested as applicable to all animals in which no nervous system has been discovered.

348. Those animals which are destitute of rays consist of a single, minute, delicate mass of jelly, covered over by or have certain spots covered with a viscid fluid, which causes nutritive particles to adhere to their surface. And the moment any organic atom comes in contact with the body of such an animal, it immediately forms an indentation ; into this the atom becomes imbedded, and the depression occasioned by it gradually becomes deeper and deeper, while the prey

sinks little by little, and causes the *zooid* (the name given to a single one-celled zoophyte), to assume a flask-like form, by a regular infolding of its margin; and thus as the edges close over the food, rudimentary rays protrude, and encircle the food so tightly, that it is entirely shut in. The enclosed particle gradually passes to the central part of the body, where its soluble parts are absorbed by the captor. In the mean time the external portion of the body recovers its pristine condition. Any indigestible residue, such as particles of shell, finds its way to the surface of the body, and is extruded by a process the converse of that by which it entered. The zooid is only a detached gemmule or offspring, such as the young sponge; yet it is evidently above the sponge in its grade of vitality. And the evolution of the zooid may be regarded as the first prelude to a digestive cavity or stomach.

349. THE STOMACH.—A *stomach*, providing it can live, is an indubitable animal. Whatever differences may exist among naturalists as to the classification of an embryo-cell, there can be no difference of opinion as to the animal character of a stomach. All appendages, all complexities of structure, are but super-additions to this one sufficient evidence of animality; and, provided it can feed itself, the stomach is alone capable of enjoying an independent existence. This cavity is in all Animals, formed by a *reflexion* of the external surface, of which the *Hydra* may be regarded as presenting the simplest example.

350. The HYDRA, or Fresh-water Polyp, exists abundantly in stagnant pools of fresh-water. Any number of them may be easily obtained in the summer-time, by simply dipping up a glass of "green-water." They re-

FIG. 30.

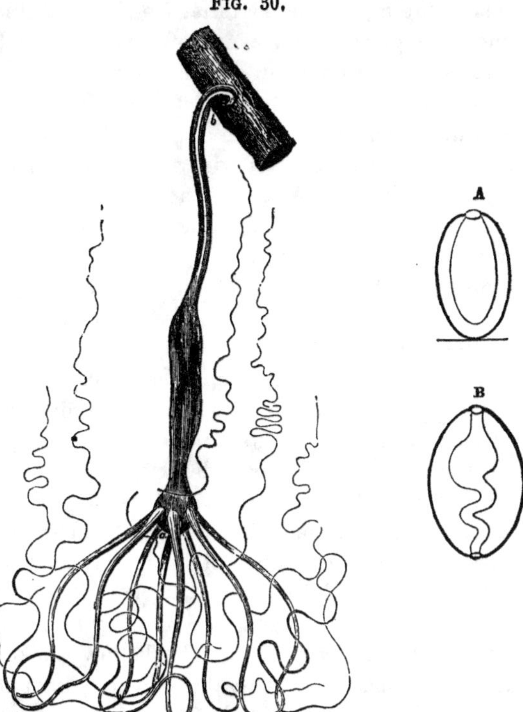

HYDRA, *or* FRESH-WATER POLYP, attached to a piece of stick, with its arms extended, as in search of prey ; *a*, the mouth surrounded by tentacula ; *b*, foot or base, with its suctorial disk. A, a stomach, with one opening ; B, a stomach, with two orifices.

semble so many little tangles of silk, of various colors, but most commonly green, and are generally about four lines or a quarter of an inch in length. The body of a Hydra consists of a single cavity, or bag-like pouch, with one opening, around which is a circle of contractile filaments or *tentacles ;* the whole substance being of a transparent gelatinous material, yet it is capable of moving from place to place, and of capturing food. The tentacles secrete on their external surface an adhesive mucus, and the base or end of the foot-stalk is provided with an adhesive vesicle or sucker, by means of which, on coming in contact with a substantial surface, it sucks itself fast. And if by accident, or otherwise, it is detached, and drifted off, no sooner does one of the tentacles come in contact with any substance than it adheres ; the arm immediately contracts, and the body of the Hydra is hauled up. By this manœuvre, the other tentacles are brought in play, and the creature is made fast. These movements, however, are all effected with great slowness, and when we consider the minute size of the Hydra, it is obvious that, to proceed a few inches in the manner described, would occupy many hours. By another means, it is capable of more rapid motion. If unfortunately cast adrift by some brisk eddy, and in danger of being wrecked or driven beyond the reach of a good harbor, it avails itself of the somerset thus brought about, by force of its adaptation to circumstances. The hollow sucker of the foot no sooner protrudes above water, than it becomes inflated ; the tentacles also being lighter than the body, the central part or disk becomes depressed or concave, while both ends are elevated, and sufficiently buoyant, to support the weight of the body. Being now converted into a *boat*, by means of its tentacles, it rows itself in the direction of food or light, with the greatest ease ! Yet the powers of the microscope have thus far been exhausted on these animals, without the vestige of a nervous system, sense of sight, or aught else, but the sense of touch, contractility, and digestion.

351. Contemplating the nature of the substance of

which the Hydra is constituted, we might presuppose their food to consist of the softest fragments of vegetable matter, but this is far from being the case. On the contrary, the Hydra captures and devours highly organized and active animals with the greatest voracity. But *how* it captures and holds such prey are interesting questions.

352. Delicate and almost invisible as are the *fishing tackle* of the Hydra, when waiting for prey, as represented in the figure, no sooner does the larva of an insect or minute crustacean come in contact with one of the outstretched lines, than instantly, as if the hand of an enchanter had been laid upon it, its movements are arrested, and all its efforts to escape rendered useless ; the active young insect, which was but a moment before, capable of darting through the water at a rate, that in a single minute, would exceed a lifetime's peregrinations of the Hydra, is now seized by the sluggard, and held beyond the possibility of escape. The tenacious filament, having taken secure hold, begins to contract slowly but surely, drags the victim towards the opening of the bag, the mouth of the Hydra. Other tentacles entwine the captured creature, as it is slowly dragged along to the open orifice ; and thus active larvæ of insects and microscopic fishes, are overpowered and swallowed by their apparently contemptible assailant

353. More than one hundred years ago,* Trembley thus graphically described the Hydra. He states that, "In the month of June, 1743, having caught a great many little fishes, about four lines in length, the first thing that I did was to try if the polyps would eat

* Trembley, *Mémoirs pour servir à l'histoire des Polypes d'eau douce*, 1744.

them. I, therefore, put several of them into the glasses where my polyps were, and experience soon taught me that, as I had at first suspected, the strength and activity of the fishes would enable them to make no small resistance; indeed, I did not dare to flatter myself that the polyps would succeed in catching them. The fishes, however, in swimming about, soon came in contact with the arms of some of the polyps, and then began furious battles, which did not always end with the same success. When a fish only touched a single filament, it was generally able to disengage itself by a violent effort, and sometimes even broke off the arm which had laid hold of it; but, if caught by several arms at once, the combat took another turn, and the fish generally had the worst of it. The struggles it made to get at liberty were, for the most part, useless, and only served to entangle it more in the arms of its enemy. Still it was easy to see that the polyp had to make great efforts to hold the fish. The tentacles which enveloped it, were much thickened, and firmly grasped the little minnow on all sides. Even when I saw a polyp had succeeded in mastering a fish, and had dragged it to its mouth, I did not understand how, by possibility, it could be swallowed. A bulky fish, four lines long, which could not be doubled up, had to be taken into the stomach of a little polyp not more than half its own length. However, the Hydra undertook the task, and accomplished it;—the tail was swallowed first, and gradually the whole fish received into the dilatable bag; which being achieved, the polyp was so tensely stretched over the included victim, that, if a person had been ignorant of the fact, he would have thought he only saw a fish, at the anterior extremity of which were a few barbs, some lines in length.

354. This little minnow, therefore, was lodged whole in the stomach of the hydra, which became, by the inordinate distention, reduced to a transparent film; nevertheless, it was soon digested. In a quarter of an hour, it was killed, sucked, and the remainder ejected from the mouth of the polyp; recognizable, indeed, as a fish, but very considerably mutilated."

355. Notwithstanding all this, the Hydra is a mere living bag, presenting no distinction of organs, no complexity of structure. And Trembley assures us, that the Hydra, if turned inside out like a glove, still has the power of seizing prey, and digesting it with the same facility! Such are the capabilities conferred upon structures by the presence of the principle of LIFE. Yet this mysterious principle always acts in accordance with certain laws. The same illustrious biographer, Trembley, informs us that, while the Hydra digests other animals with such facility, it has no power of dissolving its own substance; that when fragments of the tentacula have been broken off and swallowed with its victim, this latter is speedily digested, while the tentacle is cast out unacted upon. That, in one case, he witnessed the swallowing of a small Hydra, while in possession of prey, by a larger one, and while he expected to see it dissolved and destroyed, he only witnessed the digestion of the food which was fairly in possession of the smaller Hydra, which was immediately afterwards thrown up uninjured.

356. It is a general law of vitality, that the materials of nutrition can only be introduced into the living system in the fluid state. In virtue of this law, it becomes us to make the most of the vital phenomena, exhibited by the most simply constituted animals. As the Hydra wholly consists of an aggregation of gelatinous cells,

it is opportune here to ascertain in what manner, after the food has been digested in the stomach, the nutritive materials are appropriated to the nourishment of the fabric. Trembley, in the course of his experiments upon this subject, fed his little pets with food of different colors ; with the blood-red larvæ of certain insects, for example ; in order that, if possible, he might trace the channels through which the digested matter was absorbed. He was soon surprised to find that the whole surface of the polyp gradually changed to the same tint ; and it was only after a considerable lapse of time that the polyp regained its natural hue. This was all that could be perceived ; no connection was visible between the external surface and the stomach of the animal, neither were there perceived any channels of intercommunication between the cells which constituted it. All that could be ascertained was, that the food introduced into the cavity was acted upon mechanically by the motion of its walls, chemically by the juices secreted from its surface ; so that the nutritious parts were reduced to a liquid and absorbed, while the remainder was ejected. A hundred years' investigation by innumerable physiologists and chemists, have added nothing to the functions of the stomach When existing alone, as in the Hydra, its powers are just the same as when assisted by the digestive apparatus of the most complicated of Animals.

357. The multiplication of species in the Hydra is accompanied by gemmiparous division, in a manner similar to the same function in the sponge. But in addition to this—their whole body consisting of an aggregation of cells of the same kind—any portion is adequate to perform the needful functions for the main-

tenance of its own existence independently of the rest. If, therefore, the Hydra be cut in two, or many times divided, each fragment possesses the power of reproducing a complete animal.

358. SEA-ANEMONES.—Next to the Hydras of fresh-water marine animals of the same class, the ACTINIÆ proper, or the Sea-anemones, have attracted most attention, on account of their magnitude and the exceeding brilliancy of their colors. These, like the Hydras, essentially consist of a stomach, and nothing else. But their fleshy substance assumes a tough fibrous consistence of great strength. They are generally fixed to the rocks, or to hard bottoms in the shoal waters of sea-coasts, by a broad fleshy base, and seem to be entirely dependent upon chance bits for food.

359. Each of the long movable lips or tentacles with which the sea-anemone is provided, is hollow for the greater portion of its length, and when in want of food, filled with water. This condition enables them to float in graceful curves in all directions, like the beautiful plants after which they are named ; resembling, as they stud the bottom of the clear waters of newly formed coral islands, so many beautiful carnations or elegant bouquets. Some species are more than a foot in diameter, and arrayed as they are in the most gorgeous colors of deep red, blue, purple, and gold—when fully extended, they present a truly magnificent appearance. They

FIG. 31.

SEA-ANEMONES.

have no eyes, yet so susceptible are they of light that not a cloud can pass before the sun but they will show by shrinking that they feel the change. And, if danger threatens, they have the power of retracting the tentacles and closing up into the appearance of a rounded lump, scarcely distinguishable from the surface upon which they rest.

360. The exterior surface of the sea-anemone, like that of the hydra, when in want of food, secretes a glutinous substance, and woe be unto the crab or careless fish that comes in contact with one of these strong arms. It is "bagged" with as much certainty as the sportsman's game. A tentacle, when it has seized prey, immediately contracts; this contraction has the effect of causing an inequality of surface round the orifice, and by continuing to draw up, this inequality or indentation is increased, and other tenticles are by this means inclined in the direction of the one in possession of a captive. As each one bends its point to the aid of its fellow, water is forced from the interior, and a corresponding shortening produced. This process continuing, the food gradually becomes encircled by the whole and still shortening by degrees, they bring the food within the orifice. A sea-anemone a foot in diameter could doubtless hold and swallow a ham, or anything else as large, for they are by no means dainty in their sense of taste, whilst they are great gormandizers. Dr. Johnston, in his history of the British Zoöphites, describes an accident which happened to a sea-anemone in his possession, which in its normal state was not more than two inches in diameter, that contrived to swallow a pectin shell as large as a saucer! The poor thing managed to get the shell turned the wrong way, that is, so placed as

to divide the stomach horizontally. The little creature was now so tensely spread out as to be as flat as a pancake—the shell completely cutting off all communication between the mouth and the inferior portion of the stomach. Yet it still *lived* and kept plump, the shell seeming to have only contributed to sudden growth. On examination, it was found that what had been at first regarded as a very untoward accident, had been ingeniously taken advantage of to increase its enjoyments for the future, by providing for double fare. A new mouth, furnished with two rows of tentacles, was developed upon what had before been the base, leading to the stomach ; and the middle portion of the stomach was divided into two basal ends, each in conjunction with the shell—the animal having made use of it as a new point d'appui—and became a sort of double Siamese, but with greater intimacy and extent.

361. The sea-anemones are, indeed, so blindly voracious that anything and everything coming in contact with the tentacles is seized by them, and, if not too weighty, carried straightway into the stomach, to be there *tried* for food, and if such substance has any juice in it, it is absorbed and adds to the mass of jelly ; if not, it is ejected as refuse by the same opening through which it entered ; this constituting the whole process of digestion in these creatures. Yet they play a most important part in the economy of nature. Planted everywhere on the surface of submarine islands and in the shoal waters of dry land, they encompass the whole earth with a girdle of life, whose office it is to retrieve the organic remains of plants and animals ere they corrupt the waters, and organize them anew for an order of beings in the fathomless depths of the ocean.

362. The multiplication of species, in the actinæ is accomplished by the process of budding, or by division. In some species, denominated *Compound Polyps*, the reproduced buds or gemmæ remain attached to the parent until they produce a tree-like appearance, there being in this condition a condensation of the surface mucus which sustains them in the longitudinal growth

Fig. 32.

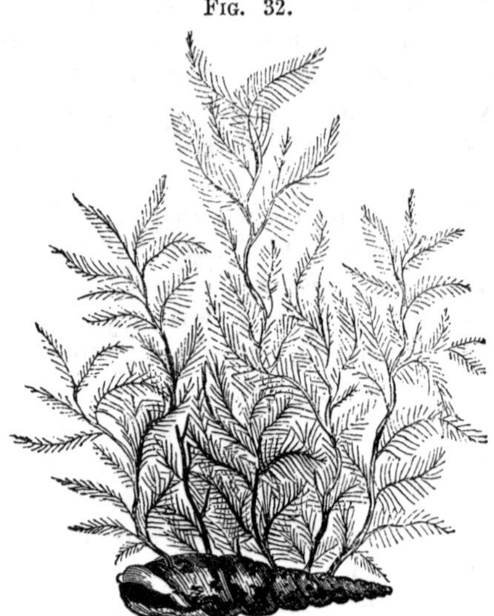

The Sertularia.

and multiplication. Thus connected externally, their digestive cavities in like manner connect with one another in such wise that a single current of water flows through an entire series of animals, one piled upon another. Through a system of tubes thus formed the contained fluid moves with considerable regularity, and in a manner which cannot be clearly accounted for, as they are not observed to contract. It is probable, however, that each individual possesses a certain degree of independent action, like the single ray of the anemone, and as each one being placed end with end contracts, the fluid is forced, in the first place, to the one below it by the closure of the oral orifice of the one at the top of the series. This process being repeated by each individual in the series, the last one at the base being provided with but a single orifice is obliged to eject the fluid and the remaining dregs by the same way it entered, and immediately close, to prevent its re-entering ; this has the effect of reversing the current, each one now being obliged to imitate the last in the series. A current of granular particles is seen moving along the axis of a series of animals thus arrayed, which, on arriving at either end, ceases a moment, and then changes in the opposite direction, and so continues flowing up and down the system, like the sap of a tree, excepting only that it continues in a digestive canal. Compound *polyps* are frequently mistaken for the *Bryopsis* (Fig. 14, p. 65), and other species of sea-weed.

363. The ACTINÆ exist in innumerable varieties of species and form—from the size of the smallest microscopic dot, up to masses in which millions of such dots are aggregated, all the progeny of a single cell! Indeed the whole *nisus* of vitality in these creatures seems to be the multiplication of species. And the larger they are, so much the more wonderfully prolific do they appear to be. The whole mass of jelly contained between the walls of the stomach and the outside film, is incessantly isolating cells, which are constantly budding forth upon the surface, or find their way into the cavity, to be thrown off as countless millions of *zoöids*

or young polyps, to become the food of multitudes of more exalted animals, which, without such sustenance, could not exist. But so numerous are they, that myriads escape being devoured, and these absorbing the soluble nutriment contained in the waters are developed, first, into the form of single tentacles, and then, fastening themselves in the midst of the organizing material of shoal waters, into beings like those which gave them birth.

364. Life in its innumerable forms is so familiar, that we treat with indifference the greatest cause—the exceeding variety in its existence—why we should admire it. The whole superficies of the land and water of the globe has been estimated at two hundred millions of square miles, seven tenths of which is occupied by water ; and a portion of the remaining three tenths actually beneath the level of the ocean. The Pacific Ocean alone being of greater extent than all the dry land upon the surface of the globe. The surface of the earth presents soils adapted to every production ; and climates which seem to be suited to every form of development. But, while the earth nourishes beings on its surface only, the whole ocean is filled with Life. Some creatures on the earth may, indeed, burrow to the depth of a few feet beneath the soil, and some may rise a little distance into the air, but that militates not against the truth, that the earth is only populated on its surface ; and the vegetable banquet there spread, is sufficiently abundant to afford sustenance to all terrestrial animals. Far different is it in the ocean. The waters are inhabited at every assignable depth ; from its profoundest recesses to the crest of its highest wave. And where, by possibility, could vegetable materials be found to feed so great a number? Food

must be supplied, and that food must be of a living kind, and in abundance to maintain myriads of hungry mouths, beyond all human estimate.

365. The bottoms of shoal waters, and the surfaces of submarine rocks, are thickly clad with vegetation. This alone, however, would afford but a scant supply, compared with the demand for nutriment in the fathomless depths of the ocean. How wonderful is the benevolence of that BEING who provides for the *transition* of *this vegetation* into the necessary food for a multitude so vast!

366. A CORAL POLYP or *Madrapore*, possesses no peculiarity of structure, neither external or internal, by which it can be distinguished from those polyps which do not secrete Coral. When the bud of a Madrapore is cast off, like that of other young polyps, by means of the ciliated tentacles with which it is provided, it for a time enjoys the privilege of motion, as if for the purpose of selecting a site; on having found one, it roots there forever. It may be a new Dover Cliff, or it may be the corner-stone of a continent.

367. When a large number of Coral Polyps are, by the force of currents, planted together, they may be seen in the clear still-waters of their growth, shooting up their brilliant stems in groves of every conceivable shade of beauty. In these living groups some of the flowers are fully open, while others are closed in the act of taking food. All are constantly changing their appearance, from bud to blossom, and from open flower to bud again. Living Coral Polyps always work upwards towards the light, and most vigorously when washed by the beating waves of a rocky bottom. They cannot exist, however, at a greater depth than about thirty fathoms, hence it is to be presumed that

all Coral beds must have been deposited within this distance from the surface. Their peculiar office seems to be to plant themselves upon submarine elevations, and build them up to the surface of the water.

368. Picture to yourself an extensive submarine plateau, carpeted with luxuriant vegetation; but every blade of grass, and every flower, instinct with Life, and all the whole expanse engaged in obtaining from the surrounding water materials for sustenance. From age to age the wide-spread scene is building up a rocky territory coextensive with itself. Gradually the accumulating pile rises towards the surface of the sea, and, at length, after the lapse of ages, portions of the rocky fabric show themselves above the waves. Here, further growth is checked; the polyps cannot live beyond the point where water freely reaches them—from whence they may derive the means of nutriment—hence, so soon as the island has reached such a height as to remain dry at low water, they leave off building higher. Yet they toil on round the edge of the fringing reef. The structure reared, becomes a nucleus round which materials are gathered. Fragments of sea-weed gather round and cling to the rough margin; dead coral, and sea-shells of every kind, strew their fragments. And the breaking surf driven by the storm tears up, pulverizes, and commingles the animal and vegetable substances, which, at every ebb of the tide, are exposed to the rays of the burning sun.

369. The heat of the sun so penetrates the mass when it is dry, that it bakes and splits in large fragments. These flakes, so separated, are raised one upon another by the waves at the time of high water, and the always-active surf continues to throw up shells, marine animals and sea-weed between and upon them.

Now peering above the waters, entire trunks of trees, which were carried into the sea by the rivers of other countries and islands, after their long wanderings, find here, at length, a resting place. And with these trees come some small animals, such as lizards and insects, as the first inhabitants. Flocks of migrating birds visit the new land, and bear with them the vegetable seeds of their previous abode. These seeds take root, spring up, and grow, and the strayed land-birds take refuge in the bushes. At last, long after the work has all been completed, Man comes—builds his hut on the fruitful soil, and calls himself lord and proprietor thereof!

370. On examining a piece of dry coral (Fig. 27, p. 130), the surface is found to be covered with numerous little depressions, hundreds, perhaps, to a single square inch. Each of these mark the place of a single madrapore or coral polyp. The coral in its live state, being *within* the polyp or series of polyps, for which it forms an internal frame-work or skeleton, analogous to the skeleton of a higher animal.

371. There are, however, two kinds of coral secretions, one within the animal, and mostly calcareous or living, and another without at the foot or base of the polyp, which is usually horny. These two secretions are sometimes mixed or applied in separate layers one enveloping the other, such as exists in the common variety known as the sea-fan, usually thought to be a sea-weed, it is coral. These two kinds are distinguished as *tissue* secretions and *foot* secretions; the former being the chief, and every portion of this during the progress of its construction is within the animal structure, constituting a common central or axial line, like the axis of a plant. The community of polyps

have their bases or secreting surfaces agglutinated to this stem while their mouths open outwardly in all directions in search of food.

372. The species of polyps which contribute to the formation of coral, are very numerous, and differ greatly from each other in the form of their habitations. The most general form is that of an irregular branching shrub, Fig. 27. But some form large round masses with numerous winding depressions, called brainstones ; some are studded with holes, filled with thin shelly plates placed perpendicularly, and converging to a point in the centre.

373. *Tubular polyps* are indicated by their name. They exist within a hard case or tube, from the apex of which the polyp protrudes in search of food, or retires within when danger threatens.

Fig. 33.

Tubipora Musica.

374. The frame-work of the *tubipora musica* consists of successive stages of cylindrical tubes, of a rich crimson color, placed one above another, until the whole fabric is not inaptly compared to the pipes in an organ, whence the origin of its name. The tubes placed upon the upper stage are inhabited by the most recent race of polyps, the occupants of those below having successively perished as new generations accumulated above them. Some other species assume the appearance of mushrooms or goblets.

375. The annexed figure is drawn from a specimen in the United States Naval Lyceum, Navy Yard, Brooklyn. It is two and a half feet high, sixteen inches in diameter, and would hold about ten gallons. This specimen was obtained at the depth of sixty feet in the Bay of Bengal.

Fig. 34.

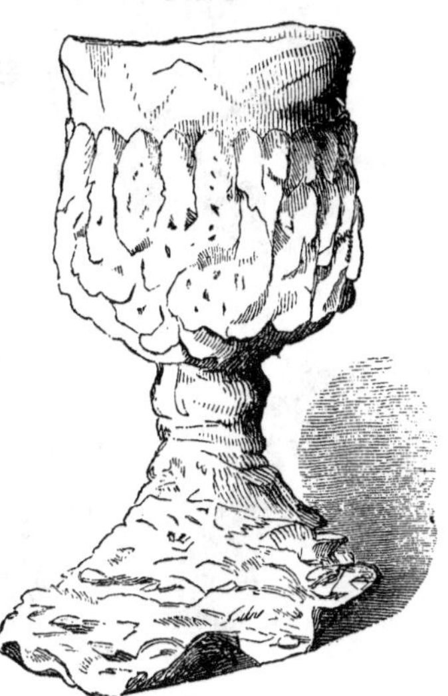

Alcyonia Gigantia—Vasa Neptunis, or Neptune's Drinking-cup.

376. The well-known coral, *corallium rubrum—coral par excellence*—so much esteemed for the manufacture of ornaments, is secreted in layers. In its living state, every branch is crusted over with a thick coating of a fleshy substance, and from the surface of this living polyps extend themselves on all sides, after the manner already described. In proportion as the stems increase in size, calcareous matter is continually deposited layer upon layer like the coats of an onion, these layers usually being colored with a rich red tinge.

377. Many other beautiful varieties of coral might be described, but these will suffice to illustrate the wonderful displays of life, in these—

"Unconscious, not unworthy, instruments,
By which a Hand invisible *is* rearing
A new creation in the secret deep."—*Montgomery*.

FIG. 35.

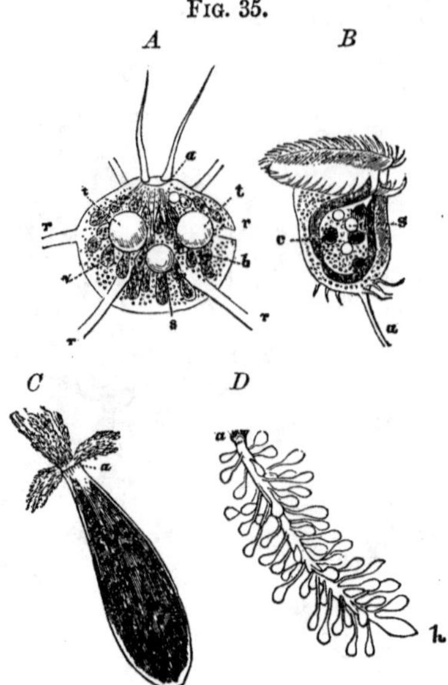

POLYGASTRIC INFUSORIA.

A, a *vegetable* monad of *valvox globator*, here shown in contrast with the animals of which it was once thought to be a species—*a*, the supposed mouth; *b*, *b*, gastric vesicles; *s*, *t*, *t*, contractile vesicles; *r*, *r*, *r*, cords of communication between the gastric vesicles or stomachs. *B*, a true polygastric or many-stomached infusoria, the *vorticella citrina*; *a*, the stalk; *s*, contractile vesicles; *t*, intestinal tube. *C*, an *enchelis pupa*, in the act of taking food, *a*, the mouth. *D*, digestive apparatus of the same animalcule; *a*, the mouth; *h*, the vent.

378. The INFUSORIA, like the actinæ, exist in innumerable varieties, but some of the higher orders of them present a higher degree of organization. Instead of a single sack-like digestive cavity, they have this many times multiplied, are *polygastric* or many stomached (*D*). The oral orifice of this digestive cavity is surrounded by a fringe of exceedingly fine rays or

cilia which appears to be the peculiar characteristic of the infusoria, distinguishing them alike from the vegetable organism (*A*), to which it has a strong superficial resemblance, and from other animals, to which they are in many respects nearly allied.

379. In constitution, the infusoria resemble the actinæ, excepting that the outer tunic in the infusoria, is more condensed and is not supplied with glutinous matter to aid it in capturing food ; this function is accomplished in the infusoria by the action of the cilia or fringe encircling the mouth. This fringe has a constant rotary motion, which has the effect of producing a current or whirl by which the particles of invisible organic matter that are constantly suspended in the water, are drawn into the mouth and admitted into the stomach or body of the cell. When colored particles of exceeding minuteness are introduced, such, for example, as the coloring particles of carmine or indigo, diffused through water containing infusoria, they are drawn in by the little eddies, and when first they enter the mouth they halt for a moment, as if to be masticated or moulded into a bolus. When a pellet is thus formed it is impelled into the general cavity, and the formation of another globular mass commences. As one after another is forced in, each, as it is introduced, pushes on the rest ; and in this way a kind of circulation of globules is occasioned, those first introduced being first to escape by the second orifice. Sometimes, however, these little animals injest particles of much larger dimensions, nearly as large as their own bodies. When any such particle is drawn into the fruitful current and comes in contact with the cilia, they contract upon it or clench it like "teeth," and, doubling in upon it, they first knead and mould it into form, and

then push it on beyond the contracted oral orifice to the interior of the sack, which expands to receive it.

380. "PHOSPHORESCENCE OF THE SEA."—A large proportion of the Infusoria display the LIGHT OF LIFE in a most remarkable manner. When a vessel plough the ocean by night in certain latitudes the waves have a luminous appearance, and if a portion of the water be taken up in a glass, and shaken, it continues to exhibit stars and ribbons of light. And if a drop of it be examined under the microscope, the source of the luminosity will be perceived to consist in an infinite number of animalcules, chiefly of the species above represented, the *Noctiluca miliaris*. It at first appears to be a minute lump of jelly, but on closer inspection it is found to consist of a sac, with definite walls, which is for the most part filled with fluid, traversed by a net-work or cellular structure of more consistent substance, containing numerous little sacs or vacuolæ, the size and form of which are incessantly changing. The light may proceed from the whole body at once, or from parts of it, as passing in succession from one to another of the sacs ; the whole function of the little creature seems to be to emit light, yet there is no organ specially adapted to this purpose. The light does not

FIG. 36.

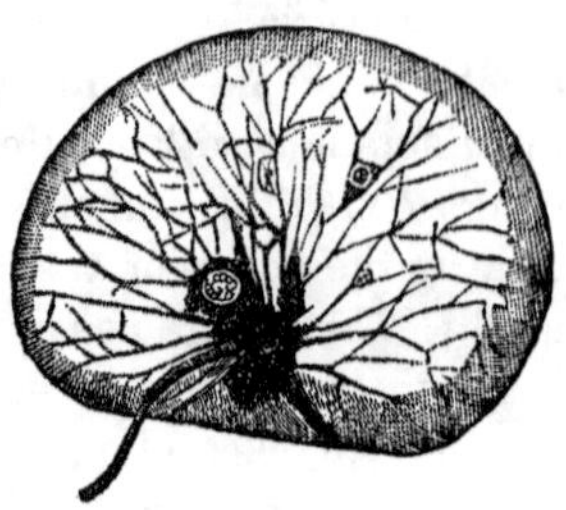

NOCTILUCA MILIARIS.

appear to depend upon any phosphorescent secretion, as no such substance has ever been discovered in connection with them. It is emitted by scintillations, which follow in such quick succession, as to cause a constant glow of light. And when we consider that a single drop of water contains hundreds, we are prepared to appreciate this wonderful phenomenon.

381. I have seen many displays of this beautiful function of animal life, but one of such transcendent beauty, and under such peculiar circumstances, as to be deemed worthy of special detail. It was after a tedious cruise from the Cape de Verde Islands, down the Coast of Africa, to about 15° S. latitude, when, having finished our course southwards, our ship was heading westwardly, and under a fine trade breeze, running at the rate of nine knots. The sun was low, and, though apparently clear, a thick haze, which hung between sea and sky, now threatened to shut out the hour which, of all others, seemed most opportune for a first sight of the honored prison, St. Helena. As the haze thickened, and night set in, sad reflections linked themselves with the great warrior, and deep gloom settled on the countenances of all on board. Presently the wind died away, but the huge black cloud now began rising rapidly out of the sea, and, save the deep-toned commands for shortening sail, silence reigned throughout the ship. Dew came down like rain, and with it a chilliness that made one forget for the time that he was still in the tropics. The "cloud" had swollen to enormous magnitude in an extremely short period of time, and had assumed a most grotesque appearance. Were it possible to contemplate an immense grotto revealing all its inconceivable dark caverns by turning inside out, one might realize the semblance. Night closed

in on the astonishing reality of a first view of St. Helena!

382. In the neighborhood of isolated islands the watery vapor of the atmosphere condenses into heavy damp clouds, which, at times, envelop the whole island, and the darkest nights in the tropics are in the vicinity of such places. Here, too, that pleasant interval of twilight suddenly merges into deep darkness—and no night ever seemed so dark as this. All at once, reflections on events long passed seized hold of one, as if to intensify the gloomy thoughts which sought sympathy with him, of whom it was attempted to place beyond the bounds of sympathy. Lost to everything immediately connected with ourselves, contemplation was just suiting itself to the thick darkness, when the attention of all hands was aroused by a succession of quick puffs and deep inspirations, as though some sea-monster were in the agonies of death. All eyes were instantly turned in the direction whence the sounds came, and there the water was alive with light. Suddenly, in the midst of it, a fountain of flame burst forth, bespangling the air in all directions. Then, up shot a blaze of light, accompanied by a deep surge-like groan, as if the great Leviathan were belching fire! But we had no time to watch the motions of the great whale as he sported in light, for our ship was dividing a foam of flame. As we looked over the bulwarks, rays like lightning darted down into unfathomable depths, and from these recurrent tangents splashed themselves into ten thousand stars, as they beat against the sides of the ship; and a long stream trailed in our wake like the tail of a comet! We were in the very midst of animals, more than a thousand of which can be held upon the thumb nail. Yet by these animals the

bases of mountains are laid, forests of coral grown, dry land built up. And when their LIGHT goes out their remains are deposited in a cushion of down at the bottom of the sea, where they form the nuptial bed of nations—the bed of the Submarine Telegraph!

CHAPTER XV.

THE CONSUMMATION OF ORGANIC DEVELOPMENT.

383. In living beings of a complex nature we discover a variety of changes resulting from the exercise of the component parts of every such being. These actions are the functions of the structures performing them, and are termed *Organic Functions.*

384. Throughout the whole Organic Kingdom, the functional character of the organs which all living beings possess in common remains the same ; but the manner in which any function is manifested varies with the general plan upon which the organism is constructed. In the lowest forms of living beings the entire fabric is made up of repetitions of the same simple elements, and every part can perform its functions independently of the rest. The same is the case in the earliest stages of development of all the higher forms of organization ; they are at first composed of a homogeneous mass of cells. Such a body, in its entirety, consists of a single organ, which possesses the power of sustenance and growth. The *zooid* is an example. It is an organization without subdivision, possessing the single function of Nutrition, and no other ; by means of this, it lives and grows.

385. Nutrition is the function of growth. It consists in the transmutation of inanimate matter to animate

matter, and comprehends all the means by which this great change is effected. Nutrition is the first function of organic existence, and the first object of a living structure ; it is therefore common to both Plants and Animals.

386. The first act of Nutrition, in the train of vital operations, is *Absorption.* Absorption signifies that process by which alimentary materials are introduced into the body. In the Protophytes, or First Plants, Absorption is performed by the entire surface, and such plants, in the exercise of this process, may be regarded as all root. It is the same case with the germ-cell of the higher plants, and in Animals ; the germ absorbs the nutriment in which it lies embedded by its whole surface. But Nutrition does not consist in growth alone ; it involves development in every change of organization ; and the conditions under which the act of Absorption is performed, differ in proportion to the heterogeneousness of the organism. And in the processes of organic development, the necessity of a *digestive cavity* for Animals is made manifest by the more or less *solid* nature of their food, and by the occurrence of intervals between the periods of obtaining it.

387. The first effort of adaptation of a homogeneous tissue to the function of digestion is made apparent by the evolution of the zooid ; and by it, is evolved the utility of a *stomach.* And in the series of animals next above the condition of the zooid a *stomach* becomes the peculiar mark of distinction. By means of this one organ, all the distinctive features of an animal body are made manifest. The stomach of the Hydra, with its single orifice, surrounded by tentacles, is to that animal the power of sensation, locomotion, ingestion,

circulation, and nutrition. In beings of higher organization each of the functions of the Hydra, all of which are performed by one organ, require a number of organs constituted of many tissues. And the more subtle the faculties of an animal, so much the more is the bodily structure complicated by different organs adapted to particular functions.

388. Among the general differences of organic beings none are more striking than those existing between the *aliments* by which they are respectively nourished, and the mode of their *ingestion* or introduction into the system. In addition to the absorbing organs with which plants are furnished, and by which they imbibe their aliment from the inorganic world, nearly all animals are provided with *cavities for the reception of food*, and for its reduction to a state fit to enter *circulatory vessels*, into which the liquid prepared by *a digestive process* in the cavity, transudes, to be conveyed to the remoter parts of the organism. The successive distinctions of animals are a *respiratory surface* through which the circulating fluid is ærified, and *secreting glands* for the separation of injurious products from this fluid ; organs of *support* and *protection*, forming a "skeleton" of some kind, either external or internal ; organs of *sensation ;* organs of *consciousness* and *self-direction ;* and organs of *locomotion.*

389. An *Organ* is the aggregate of a number of tissues, of various kinds, and every organ is made up of a multitude of cells, each one of which is adapted to the exercise of functions, even as the organ they constitute is adapted to its functions for a higher end. The collection of organs which concur in any function, is called an *apparatus.*

390. The exceeding small size, and the countless

number of pieces with which Nature works in the creation of organic bodies, ensures an adaptation to particular purposes, wholly unattainable by artificial means, and it is only by comparing Nature with herself that we can appreciate the realities of her infinity. Light is known to travel one thousand millions of feet in a second, and the length of each separate undulation of light averages one fifty-thousandth part of an inch, yet a ray of light which proceeds from a star of the twelfth magnitude requires not less than four thousand years to reach the earth. Estimates like this may be applied to the enumeration of the cells which go to make up a single organ or the entire organism of an organic body. The breathing tubes of a rabbit have been estimated to sustain twenty millions of cilia—the special organ of the Infusoria, each one of which cilia is constituted of an aggregation of cells. And the breathing tubes of a man sustain more than one hundred and fifty millions of cilia, every one of which is in perpetual movement! The root of a single nerve of the human brain contains more than a hundred thousand primitive fibres, and each of these fibres is constituted of a number of cells. And the last molecules which are revealed by the aid of the microscope, are all made up of smaller cells, and these are the material of Nature's laboratory for every organic structure.

391. The first necessity for organic development is the circulation of a nutritious fluid. In the protozoa and nearly allied orders of animals, the nutritious fluid or "blood," as it is incorrectly called, is nothing but water, like the sap of the plant, containing nutritive particles. And as such animals consist of a homogeneous tissue throughout, the fluid which they absorb

pervades the whole tissue alike as it does in the thallus of the sea-weed. The evolution of the circulating system in animals conforms in all essential respects to that which has been described in the vegetable kingdom. But in proportion as the power of absorbing aliment is restricted to one part of the surface, whether external or internal, it becomes necessary that means should be provided for conveying the nutritive fluid to distant parts of the organism.

392. There is, however, one important distinction between the circulating apparatus of plants and animals. In the plant, the sap ascends to the leaves and is there ærified and *elaborated*, or undergoes certain changes which result in the formation of new and peculiar products termed the *proper juice* of the plant. The remaining liquid from which the juice is elaborated, is then exhaled or given off by the upper surfaces of the leaves. Meanwhile, a system of vessels on the under surface of the leaves takes up the proper juice and distributes it throughout the fabric, giving up its nutritive constituents to the formation and nourishment of the various tissues of the plant. In the higher animals, on the contrary, the *same* fluid is *repeatedly* transmitted through the body, the alterations which are effected in one part of its course by the withdrawal of nutritive constituents, being counterbalanced in others by different functions, chiefly by those of respiration and secretion, and by the continual admixture of new alimentary materials.

393. As in the simplest beings the entire surface participates equally in the act of absorption, so in the most complex beings, every part of the surface retains some capacity for it; since, even in the highest plants and animals, the common external covering admits of

the passage of fluid into the interior of the system, especially when the supply afforded by the usual channel is deficient.

394. During the early stages of development all animals depend upon absorption. The development of the organic functions in animals is so nearly alike in the earlier periods of their existence, that the bird's egg may be taken as an example of the whole.

395. The germ-cell of the mammalian ovum is at first surrounded by a mass of nutrient matter, chiefly composed of albumen and oily particles, and known by the name of vitellus or yolk, and the whole is inclosed within an envelope termed vitelline membrane, or yolk-sac. This membrane consists of three layers, the middle one of which is termed the germinal membrane, and as it gives origin to the circulation, it is called the vascular layer. In the first stages of development the thickened portion of pabulum that surrounds the germ-cell becomes studded with numerous irregular points or cells of yellowish color, undergoing transformation

Fig. 37.

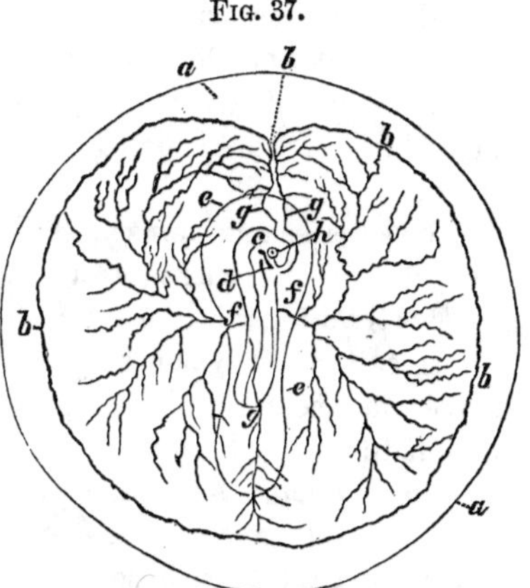

Vascular Area of a Bird's Egg.

a, a, yolk; *b, b, b, b,* venous sinus bounding the area; *c,* aorta; *d,* incipient heart; *e, e,* area pellucida; *f, f,* arteries of the vascular area; *g, g,* veins; *h,* eye.

into blood-cells ; the blood being formed before either blood-vessels or heart. Portions of these blood-cells group themselves together like so many disks, or as the cells in the yeast plant; the sides coallesce and elongate into lines, like the ducts of plants or the tubes of the uni-celled polyp, and through these may be seen coursing small blood-disks. As the lines extend themselves, the ends coallesce and form an intricate network of delicate cell-tubes. A definite plan of development, however, is perceptible from the very first. The blood-disks first array themselves into pairs, the lines follow suit and develop in various degrees of quickness, though they are all in the earliest stages of the same character, there being no distinction of arteries or veins.

396. The arteries and veins are first distinguished by the direction of the currents of blood circulating in them. Shortly after this the arteries are distinguished by their thicker walls, when they are seen distributing themselves for the formation of particular organs, following out the typical plan of the organism they are destined to develop. After the principle vessels are formed, the development of new ones are formed by prolongations of the extremities of those already existing, or by the coallescence of different branches. Meanwhile, the formation of the blood goes on in every part of the body, and, when formed, it is impelled by some unknown cause in the right direction, until at length it reaches a central point in the area which is first formed by an aggregation of blood-disks. The innermost disks of this cluster soon break down into small cells, while the outer agglutinate and form a tubular cavity, to be further modified and changed into a HEART. The first motions of the blood are *towards*

the subsequent situation of the heart, and consequently the first vessels formed are *veins*.

397. The heart at first consists of a single cavity extending nearly the whole length of the embryo, the posterior extremity receiving the veins, and the anterior extremity communicating with the arteries, a condition which is permanent in the *salpidæ* and some other molusca.

Fig. 38.

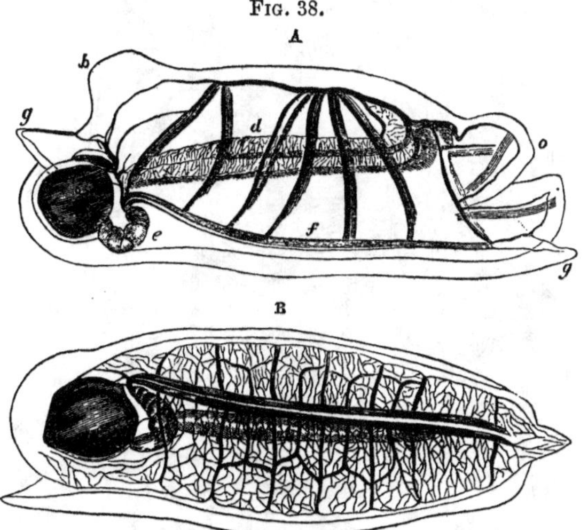

Circulating Apparatus of the Salpa Maxima.

A, as seen from the side ; *B*, as seen from the ventral surface ; *o*, oral orifice ; *b*, vent ; *c*, nucleus composed of stomach, liver, etc. ; *d*, bronchial lamina or ærifying apparatus ; *e*, heart, from which proceeds the longitudinal vessel *f*, sending transverse branches across the body ; *g*, *g*, projecting parts of the external tunic, serving to unite the different individuals into a chain, it being a habit of these animals to link themselves together and float along in the open sea. The salpa consists of an elastic external membrane which is *transparent* and open at both extremities, affording a remarkable facility for study in the circulating apparatus as here drawn. They have a complete system of capillary vessels, adapted to the diffusion of the blood over the membrane lining the general cavity, and also to the bronchial lamina, while the heart remains in the most rudimentary condition. And by the complete intercommunication of the numerous blood-vessels communicating with longitudinal trunks, the blood is perfectly circulated and ærified.

Fig. 39.

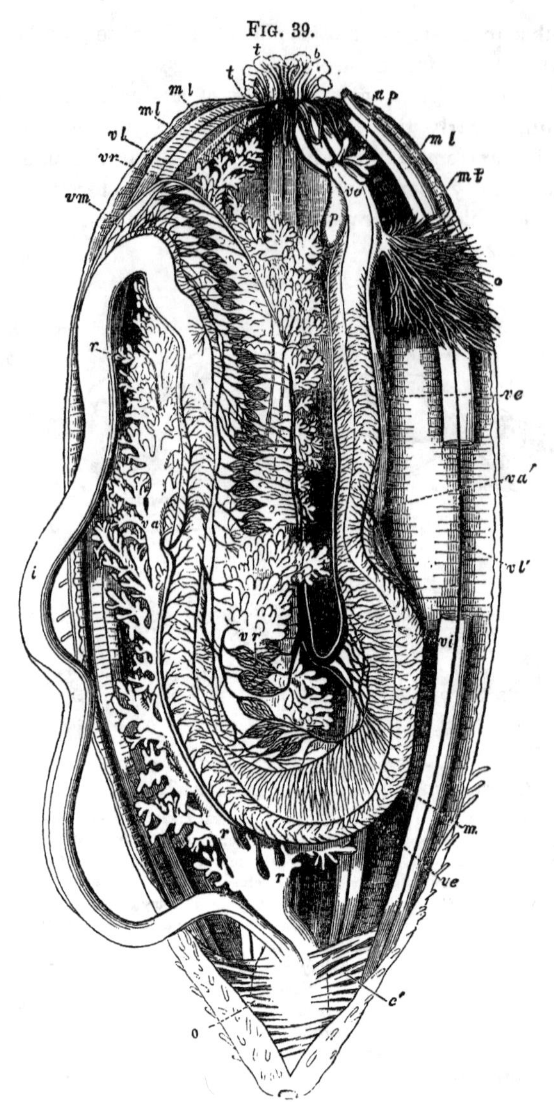

HOLOTHURIA TUBULOSA, OR SEA-CUCUMBER.

The anatomy of the Holothuria presents a singularly rudimentary condition of all the organs, in an apparently transition state between the lower and the higher animals. *b*, the mouth surrounded by tentacles ; *c*, cloaca, surrounded by muscles *c'* ; *i*, intestinal tube ; *m*, mesentery or sustaining membrane ; *ml*, *ml*, longitudinal muscles ; *mt*, transverse muscles lining the entire inner surface of the integument ; *o*, ovary ; *ap*, appendix vermiformis ; *p*, heart ; *r*, *r*, respiratory apparatus or lungs ; *t*, *t*, oral tentacles ; *va*, blood-vessels surrounding the mouth and supplying the tentacles ; *vc*, anterior portion of the intestine giving off a branch *va'* which communicates with another part of the same tube ; *vi*, another branch of the intestine ; *vl*, *vl'*, longitudinal and transverse blood-vessels of the integument or cellular tissue between the muscles ; *vm*, blood-vessels of the mesentery, connecting the branches of the external and internal vessels with those of the respiratory system *vr*.

398. The *Holothuria tubulosa*, or sea-cucumber, exhibits that degree of development wherein the heart is contracted to a single pulsating vesicle, communicating with an intricate set of blood-vessels ramifying all the tissues ; and together with this circulatory system, a corresponding degree of development in the other systems of organic life. The sea-cucumber is in some respects allied to the infusoria ; its mouth being surrounded by rays which under certain circumstances are organs of locomotion. Its movements, however, are usually accomplished by flexures of the body, and for this purpose, a series of muscular bands is provided, similar to what exists in worms. The alimentary canal is a tube of nearly uniform diameter, without distinction of parts, and may therefore be regarded as an elongated stomach ; this is held in its place by a well-formed *mesentery*, or sustaining membrane, upon which the blood-vessels are minutely distributed, and from which branches are given off to connect the external and internal vessels. For respiration there are two special provisions ; the fluid of the circulatory vessels being ærated by transmission to the tentacles, whilst

the circulatory fluid of the general cavity is ærated by the water introduced through the respiratory tree. These organs, however, all exist in a rudimentary state, and so little is this animal dependent upon them for the performance of its functions, that it is capable of existence without them. The holothuria sometimes disgorges itself of stomach, circulatory organs, respiratory tree, and all, reduces itself to a simple sack, and occasionally it throws off its tentacles; finally, it divides itself into two or more parts, each part ultimately developing itself into a perfect animal, possessing organs!

399. When the heart first begins to pulsate in the embryo-bird, it contains only colorless fluid, mixed with a few blood globules. A movement of the dark blood in the circumference of the vascular area is at the same time perceived; but this is independent of the contraction of the heart. And it is not until a subsequent period that such a communication is established between the heart and the distant vessels, that the dark fluid contained in them arrives at the central cavity and is propelled by its pulsations.

400. The only necessary function for the maintenance of life in the animal is the same as that for the plant, nutrition, and upon this depends the development of other functions; and upon the development of other functions depends the development of organs for their maintenance—functions in all cases, in their origin taking precedence of organs. Absorption of fluid being the preparatory step to the circulation of the fluid in vessels, these are developed as above shown, and as the vessels distribute the nutritive fluid to distant parts of the body, other organs are developed from the blood in the same manner as the vessels were at first devel-

oped, namely, by a coalition of blood-disks. And thus it is, that in the process of development, one organ after another is added, secreted, as it were, from the excess of nutritive fluid in circulation, until the special attributes of the germ-cell are fully attained.

401. The development of the liver in the egg of the bird will serve to illustrate the process by which every organ is formed. The first rudiment of the liver is formed by a collection of blood-disks in immediate contact with the heart round an everted portion of the digestive tube, which subsequently forms the channel of communication between the liver and the tube. The increase of the organ takes place by a continual budding forth of cells from its peripheral portion. Finally, the cells of the exterior are metamorphosed into a hardened investing tissue, while those of the interior break down into the formation of the necessary ducts.

FIG. 40.

DEVELOPMENT OF THE LIVER.

a, heart ; *b*, intestine ; *c*, everted portion becoming connected with the liver, *d*.

402. Shortly after the heart in the bird attains the contour of a single pulsating cavity, as illustrated by the permanent condition of that organ in the sea-cucumber, it contracts upon itself like the walls of the simple cell, and, finally, by the coalescence of blood-disks, forms a complete septum, dividing the heart into two cavities, named *ventricle* and *auricle*. This condition of the heart is permanent to fishes, the ventrical communicating with the arteries, and the auricle with the veins.

403. The next step in the progressive development of the heart in birds and mammals is, it doubles upon itself, the convex curve becoming the apex or point of the heart, while the length is much lessened. Next, the two cavities begin contracting upon themselves, each dividing into *two* ventricles and *two* auricles, the division of the ventricle, however, considerably preceding the division of the auricle. The division of the auricle goes on more slowly. A large round hole called *foramen ovale*, continuing for some time at the lower part of the septum, and through this the ventricle continues to communicate with both auricles. The *foramen ovale* is afterwards closed by the development of a curtain or valve, which covers it, when the division of the heart is complete. The septum between the auricles being continuous with that between the ventricles, the heart may be regarded as a double organ. The cavities of the *right* or venous side of the heart having no direct communication with those of the left or arterial side. And the vessels proceeding from these two sides are entirely distinct from each other, having no communication except at their extremities.

404. During the various degrees of the development of the Heart, there is a corresponding development and distribution of the Arteries and Veins, and, indeed, of all the organs of the system. And all the functions become so completely linked together, that none can be suspended without the cessation of the rest.

405. The Nutrition of all the tissues depends upon Absorption of ingested aliment, and this cannot be supplied without Digestion. The nutrient material, separated by the process of Digestion, requires the function of Circulation to apply it, while Respiration and Secretion are equally necessary for its purification. These

several functions acquire their greatest perfection in man, and in him their exercise depends upon a nervous system—this being a special attribute of animals. The lowest orders of animals, however, exhibit no traces of a nervous system or nervous influence of any kind ; contractility and irritability in them being of the same character as that which exists in certain plants.

406. In the progressive development of organic bodies, the several parts which constitute the entire fabric are characterized by specialties of conformation, and each part becomes a distinct organ adapted to the performance of a particular function, different from that which can be discharged by any other part. The end of all organization is perfection. And, in its course, the most perfect of its kind is when the vital susceptibility is the most intense and unhindered. The organic being is the most complete when all the forces of vitality are the most freely active.

407. Having now considered the elementary forms of all organic beings, and the conditions on which they depend, we are prepared to consider the special conditions for the consummation of corporeal organization as applied to the erect stature and expressive countenance of MAN.

CHAPTER XVI.

THE FUNCTIONS OF THE ORGANS.

408. All the conditions of life detailed in the preceding pages apply to man. The internal structure and vital phenomena of the human system, closely resemble those of the lower animals. Like them, his body consists of a collection of fluids and solids ; the fluids in the human body constitute about five sixths of its entire weight. It is from the fluids of the system that all the organs receive their nourishment and repair their waste. The solids are all formed from the fluids, and when they have served their purpose they return to their former state, and are decomposed ; solidity in man being, as in other organized and living matter, only a transient condition. Of all the animal fluids, water forms the chief part. It contains all the other essential substances in solution, and from it they are applied to the building up and nutrition of all the tissues of the body.

409. By *tissues*, is meant the various parts which, by their union, form the organs ; and the solid parts, or organs, are classified into different systems, according to the particular function which they exercise.

410. Cellular Tissue.—Of all the organic tissues which enter into the composition of the animal body, the cellular tissue is the most common and the most extensive. It is uninterruptedly distributed all over

the body, and in this manner it constitutes the common bond of union between the different regions and organs. By its elasticity and contractility it rounds off and gives shape to the various parts, and by the fluid contained in its cells it greatly facilitates the motions of parts on each other. It is accumulated in all places where there is extensive motion, as about the eyes and cheeks, in the joints, and around most of the internal organs.

411. The texture of the cellular tissue is almost described in its name ; that is to say, it is composed of a collection of air-cells or areola held together by a continuous expanse of thin membrane, which connects them all together, and spreads over and around all the organs, filling up the interstices between them in such a manner as to be endless in its distribution. It is, therefore, of variable thickness in various parts, depending upon the degree of motion and delicacy of the organ to be protected. In some parts it is spread out with the tenuity of the cobweb, as between the muscles, and over the smooth surface of the eyeball ; while in others, as in the palm of the hand or sole of the foot, it is thick and tough, and not inappropriately called whitleather.

412. Sometimes the cellular tissue becomes loaded with *fat*, and it is then distinguished as *adipose tissue* (from *adeps*, fat), in this case it forms a reservoir of nourishment, the fat which is contained in its cells being absorbed when food is withheld. This is strikingly illustrated in hibernating animals, which are very fat when they fall into the torpid state, but at the return of spring they are much reduced in bulk.

413. Besides these special purposes of the cellular tissue, the other membranes of the body chiefly consist

of it in various degrees of condensation, and it also enters largely into the composition of all the organic solids, many of which are, in fact, entirely formed of cellular tissue, variously modified and disposed.

414. The Muscular System consists of a great number of bodies called *muscles*, from *mus*, a rat, because the ancients compared the muscles to flayed rats. The muscles constitute what, in common language, is called the flesh of animals, and they occupy a large proportion of the whole body, and by their bulk they constitute a great part of the weight.

415. The muscles consist of bundles of filaments called fibres, held together by cellular tissue. It is well known that in man and most animals, the muscles are of a reddish-brown color. This color, however, is not an essential property, as it is capable of being removed by repeated washings. And some animals, birds, for instance, have some of the muscles of deep-red color, while others are white. The most important and interesting property of the muscles is that in virtue of which they have the power of contracting, or shortening themselves on the application of a stimulus, and by which they produce the various movements so strongly characteristic of animal existence. It has been well observed that the first sensible operation of life is muscular motion, and that the numerous combinations of this motive faculty sustain and carry on the multiplied functions of the largest animals, so that its temporary cessation cause the suspension of the living powers, and its total quiescence—death.

416. Muscles have been divided into *voluntary* and *involuntary* muscles. The former are those which execute movements under the influence of the will, as the muscles of the hand in the exercise of writing ; of the

head, trunk, etc., which we are capable of exercising at pleasure. The involuntary muscles are those which act under the influences of certain special stimuli, independent of the will, as the heart, muscles of the digestive apparatus, etc. There are also *mixed* muscles, those which partake in a measure of both voluntary and involuntary muscles ; of such are the muscles concerned in respiration, which we can partially control.

417. The muscles are very numerous, and are distinguished according to their situation, shape, or action. And they are grouped together as *associates* and *antagonists*, the former act together, and by their combination cause the same kind of motion ; the latter, on the other hand, act inversely to each other, and produce dissimilar movements.

418. Nearly all the various movements of the body are accomplished by the combined action of two or more muscles. And it is frequently the case that muscles which are opposed to each other in their single actions associate to produce an intermediate movement. The rotary motion of the eye and the mechanism of climbing, jumping, etc., are examples of associated muscular action to produce intermediate movements.

419. The velocity and variety of muscular movements are almost incomprehensible. An approximation has been made by an effort to enumerate the different muscular movements, produced by seven pairs of little muscles in speaking; supposing each one of these to be capable of acting separately and in combination, the number of movements is computed at no less than sixteen thousand three hundred and eighty-three ! And the velocity of these acts may be conceived of when we take into consideration the rapidity of enunciation ;

or the motions of the fingers when playing upon a piano.

420. The *duration* of muscular action is limited within certain bounds, so that after it has continued a time relaxation necessarily succeeds. Common experience proves that, in performing any laborious or continued movements, we are at length compelled to desist, and wait until the muscles are recruited before exertion can be renewed. In general, it appears that the period of rest which is required is proportioned to the degree of exhaustion ; thus, in some of the involuntary muscles, as the heart, although the cessations are momentary, they are sufficient for the restoration of power to the muscular fibres.

421. The duration of voluntary muscular action is, to a certain extent, regulated by the will, being proportioned to the degree of stimulus communicated through the nerves. Under certain circumstances the voluntary muscles are capable of sustaining continued action, according to the necessities of the individual. They are also capable of extraordinary *efforts* in special manifestations of strength, which may become necessary in overcoming resistance, or in the performance of extraordinary labors. There is, however, in all such cases a proportionate degree of relaxation subsequently required, in order that strength may be restored ; and if extraordinary efforts are too frequently exerted, the strength of the muscles becomes impaired.

422. Finally, the muscles are so placed as to be models of mechanism. Their direction and attachments to the bones furnish the archetypes for all the powers used in mechanical ingenuity. But, as they are necessarily so adjusted as to occupy the least possible space compatible with the greatest strength, there

frequently appears to be a great waste of power; this circumstance only serves the better to exemplify their wonderful adaptation to the purposes for which they are designed. For example, the deltoid muscle, which forms the fleshy part of the shoulder, is attached to the arm-bone midway between the shoulder joint and elbow, forming a lever of the third kind; it has been estimated that this muscle employs a force equal to two thousand five hundred and sixty-eight pounds to overcome a resistance of only fifty pounds. Had it been attached further down less power would have been necessary, but the contour, beauty, and adaptation of all the other parts of the limb would have been forfeited. In applying the laws of mechanics to the animal frame, the muscle constitutes the power; the joint, the fulcrum; and the bone, the lever.

423. THE BONES consist of those hard parts of the animal frame that form the *osseous system* or *skeleton.* They are the support of the soft parts, and receive the attachments of the muscles with which they combine in executing the power of motion.

424. There is no part of the organism that presents a more striking illustration of the adaptation of parts to their particular uses than the bones. This is apparent in the osseous system of all animals.

425. Many species of the lower animals and insects, whose soft parts are necessarily fragile on account of the nature of the food which nourishes them, are, nevertheless, subject to violence in wind and water, requiring for them an armor capable of resisting external injury, and yet adaptable to the circumstances of their existence. Hence we find these animals covered with shells and other integuments of bony hardness equal to the conditions of their being. These

hard envelopes are denominated external skeletons, and they are capable of being renovated or changed whenever the growth of the animal, or other cause, renders it necessary.

426. Thus the lobster or crab, changes its skeleton every year, when the body has so much increased in size as to become too large for its former covering. Previous to changing their skeletons these creatures seem to grow feeble, lose their appetite, and pine away, as we might imagine a hearty young girl to do if subjected to the constant pressure of tight stays. The muscles become flabby, the limbs shrink, and the poor creature crawls away into some secluded spot and takes to its bed ; it eats no solid food, is incapable of going abroad, and has but little appetite for the thin infusions provided for its nourishment ; restlessness comes on, and, turning on its back, the limbs are agitated and rub against each other, as if writhing in pain; at last strength is exhausted, and the quietude of sleep supervenes. After a time, the feeble arms gradually fold themselves together, and again spread out, as if seeking relief from lassitude; presently these motions succeed each other at shorter intervals ; the wasted body now realizes relief from the heretofore tightly fitting corset, and this latter, having lost its support and its use, has also suffered for want of repair. Appetite gradually returns, the invalid taking larger quantities of thin food, rapidly grows fat, and, as the body swells out, the lacing is rent in twain—now no longer fit for its purpose—the limbs are withdrawn, and the whole investment rapidly cast off. A quick convalescence ensues—appetite increases, the vital functions grow active, the body expands, and while in this condition a loosely fitting envelope covers the whole body

and limbs. But this loose covering is gradually being drawn in at one place and let out at another, to fit the now no longer restrained movements; meanwhile, this new tissue hardens, a bony substance fills its interstices, and finally it adapts itself to the habitudes of the wearer—differs from the old suit in nothing excepting that it is of larger size; allows freedom of motion for the whole organism and vigorous health is restored.

427. In man and in the higher animals, the skeleton is internal. And the bones, in virtue of their organization, are developed from the blood, and, by a process of growth similar to that of the soft parts, gradually enlarge and attain their full perfection at the time of maturity for the whole body. In these, too, the bones are constructed with special reference to the animal's peculiar mode of existence.

428. In birds, for example, the principle bones are hollow, by which their size and strength is much increased, without increasing their weight, while with these bony cavities there are air-tubes communicating with the lungs. Thus combining strength, lightness, and buoyancy suited to the habits of the animal. The whale affords another instance of the great perfection of the internal skeleton. This animal, though an inhabitant of the sea, breathes by means of lungs, and in this respect differs from fishes. Consequently, it is compelled to rise from time to time, in order to respire atmosphere. To facilitate this ascent, the bones of the head are so constructed that the windpipe, instead of communicating with the nostrils placed at the extremity of the muzzle, as in other animals, communicates with the external air at the top of the head, which, by a peculiar projection, is the very highest part of the animal when horizontal, so that it can breathe when

none of its body is exposed above the surface of the water. They usually breathe when the "*blow-holes*" (the orifices of the windpipe) are a little under the surface, and it is the force with which the air is expired in this situation that blows up the water lying over them, and which constitutes the "spoutings" of the whale.

429. If we extend our examination to the individual parts of the human skeleton, we shall be struck with still more marked evidences of excellence and design in the form and arrangement of all the bones. But where everything is so perfect, it is difficult to make a selection. What, for instance, can be more admirable than the construction of the skull? It possesses the globular form, which is of all others that best adapted to resist external violence. Its various bones are fitted to each other with the nicest exactness, and with the truest relations to the best principles of architecture. And that the brain may be protected in the most effectual manner, those parts of the skull most exposed to injury are strengthened by increased thickness, resisting bars and projecting spines.

430. In every part of the body the bones protect and support the soft parts. In the limbs they afford columns against which the muscles, vessels, and nerves are firmly sustained. In the vertebral column and trunk, they form cavities for the protection of all those parts of the organism which are most immediately necessary to life.

431. The Cutaneous System is that which forms the general envelope of the body, and is continuous with the mucous membranes, through the different natural apertures. The organization of the animals placed lowest in the scale, display in a striking manner, the

resemblance of the external and internal coverings of the body. In these, the skin and the mucous membrane are indeed so identical, that one may be substituted for the other, as shown by turning the polypus inside out. There are, however, in the higher animals, great differences between what may be called the external skin or investment, and the internal or lining membranes.

Fig. 41.

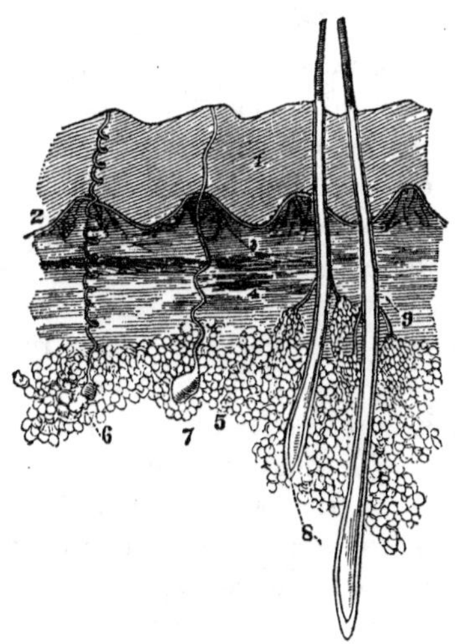

The Cutis, or Skin.

1, the cuticle or epidermis; 2, rete mucosum or mucous net-work; 3, two of the quadrilateral papillary masses, such as are seen in the palm of the hand or sole of the foot; 4, the deep layer of the cutis, the corium or derma; 5, adipose or fatty vessicles, in immediate contact with the under surface of the skin; 6, perspiratory gland with its spiral duct, such as exist in the palm of the hand or sole of the foot; 7, another perspiratory gland with a straighter duct, such as exist on the scalp; 8, two hairs from the scalp, inclosed at their roots in little bags or follicles which regulate their form of growth; 9, a pair of sebaceous or oil glands, which open by a duct into the follicles of the hair, and supply it with the oil that gives it glossiness.

432. THE SKIN consists of three layers. 1. The *Cuticle* or *Epidermis*, the outer layer, which is a kind of secretion possessing no sensibility, nor even life. It is a mere varnish-like covering, semi-transparent, and moulded over the whole surface, and on parts of the body not fully exposed to the air it remains soft and pliable, and can scarcely be distinguished as a separate layer. But on other parts it becomes dry and hard, and it is that which exfoliates as dandruff from the scalp. And the thin pellicle which raises as a blister, on the application of an irritating plaster or hot water, wholly consists of the insensible cuticle or epidermis. In lobsters and other crustacea, the cuticle becomes solidified into an opaque limy covering, but in other animals living in the water it remains soft and transparent.

433. 2. The *Rete mucosum* or mucous net-work, is that which we see when we look at the surface, to which the skin owes its color and varied tint in different individuals and races. It is of a soft pulpy consistence, delicately spread over the numerous sensitive papillæ, which are situated in the next or third layer, the *Corium, Derma,* or true skin, which is the base and support of the other layers. The corium is composed of a great number of lines of condensed cells or fibres, interlaced and crossed in every possible direction calculated to give elasticity and resistance. The skin defends the body from the injurious influence of external agents. And for this purpose all animals are provided with such a covering, although its structure and appearance are liable to great variation.

434. The appendages of the skin are the nails, hairs, sebaceous glands, and perspiratory ducts or pores. The nails are composed of numerous layers of the insensible epidermis, disposed in such a manner as to

protect the sensitive extremities in the exercise of touch. The hairs consist of condensed cellular tissue, resembling the pith of a plant. The porcupine's "quill" is a *hair* of large dimensions, and exhibits the precise structure of those growing on our heads. Each hair consists of a root placed under the skin, and of a stalk which projects beyond the external surface; they therefore serve to connect together all the layers of the skin, and these, with the underlying cellular tissue, like so many fine pins or fastenings. They also afford, in the lower animals, at least, protection of the body from the vicissitudes of temperature and moisture. The sebaceous glands (from *sebum*, suet), secrete a material having some analogy to suet, which constantly bedews the external surface, so as to preserve the softness and smoothness which are required for the proper exercise of its functions. This secretion repels aqueous fluids, as may be readily observed by placing the hand in water. It also prevents in a great degree the injury which would otherwise result from the friction of contiguous parts.

435. The perspiratory ducts and glands or pores, fulfill the important office of regulating the temperature of the body. These glands are diffused in varying proportions over the entire cutaneous surface. It has been estimated that over three thousand five hundred of them exist in a square inch of the surface of the palm of the hand; and since every duct or *tube*, when straightened out, is about a quarter of an inch long, it follows that there exists a length of tube equal to nearly nine hundred inches, or about seventy-four feet in every square inch of the skin of the palm. The average number of these pores to each square inch of the surface of the body is about two thousand eight

hundred. As the number of square inches of surface in a man of ordinary size is about two thousand five hundred, the total number of perspiratory pores may be stated at *seven millions*, and the length of these collectively is about one hundred and forty-six thousand feet or a little over twenty-eight miles. The amount of fluid carried off by cutaneous transpiration through the perspiratory pores chiefly depends upon the temperature of the surrounding atmosphere. Cutaneous transpiration is usually divided into *sensible* and *insensible*—the latter being constant, the former occasional. When the external temperature is very high, a large amount of fluid is given off by the skin, and this in evaporating carries off heat, which would otherwise raise the temperature of the body. If the atmosphere be hot and dry, and also in motion, both exhalation from the glands of the skin and evaporation from the surface go on with great rapidity.

436. *Evaporation* takes place from the cutaneous surface under the same circumstances as from an inorganic body ; and the more readily watery vapor can be taken up in the atmosphere, the more evaporation will there be from the surface of the body. In cold weather very little vapor is carried off from the skin even though the air be dry, and a warm atmosphere charged with dampness will be quite as ineffectual.

437. If the atmosphere be *cold*, cutaneous transpiration and evaporation are both checked, the former almost entirely ; but if the air is dry, evaporation continues, though to a slight degree. On the other hand, if the atmosphere be *hot*, and is saturated with moisture, the evaporation will cease, while exhalation by the glands will go on.

438. We learn from these facts the great importance

of not suddenly checking cutaneous exhalation by exposure to cold, when the secretion is actively going on, since a great disturbance of the circulation will ensue in consequence of the detention of matters in the blood which should be exhaled by the skin, as well as by the change in the temperature of the body.

439. Cutaneous exhalation is frequently ten times greater in dry than in moist air, and is doubled in passing from 32° to 68° of temperature. Other things being equal, vitality is best promoted when the surrounding temperature most nearly accords with that of the animal occupying it, though the degree of temperature which appears most conducive to the health of man is considerably below the temperature of his body. This apparent incongruity is doubtless owing to man's temperature being such as to best accord with his adaptability to all climates. This adaptability of man establishes his cosmopolite nature, and it is by this that he becomes moulded to the various climates of the earth.

440. The degree of temperature most conducive to health varies in different climates and seasons, and is also influenced by the habitudes of individuals. It may, however, be said to vary from 65° to 80°, while the temperature of our bodies is about 98°, and we are incapable of sustaining any influences which will cause a departure from this of more than ten or twelve degrees.

441. Although our organs possess a compensating power of calorification when the heat of the atmosphere is greater than that of the *blood*, so that the temperature of the body is not changed by the surrounding medium, yet we are so constituted as to make a constant abstraction of heat from the body natural and

necessary to health; consequently, if from any cause the system suspends its power of resistance to external influences, and greater external heat than that of the body produces its ordinary effects of radiation and conduction as it does on inanimate objects, so as to *raise* animal heat; or if the abstraction of heat becomes excessive, so that the body cannot regenerate it as rapidly as it is expended, and, consequently, *lowers* the temperature—serious derangement and death must speedily follow, unless the system is supported by external appliances.

442. *Variations* of temperament are far more injurious than either permanent heat or permanent cold, and these are the most common exciting causes of disease. They not only involve the sudden checking of a previously abundant cutaneous exhalation, or the converse, but all the vital functions are subject to such irregularities as, if frequently occurring, ultimately forfeits that harmonious action which is essential to health.

443. In changes from a higher to a lower temperature, or of climates involving those conditions, excessive appetite is a usual result, and while the system is likely to be surfeited, diseases are not only more likely to occur on account of the change, but they are subject to being aggravated by over indulgence. And it is too apt to be the case under these circumstances, that individuals place a false estimate on their feelings and indulge in alcoholic stimulants. These excite the system to an increased susceptibility to the influence of change, and either by the proportionate debility following the excitation caused by the stimulant, or its excessive action if persisted in, such persons are always most likely to suffer by changes of temperature or climate.

444. In cold and temperate climates, much labor is devoted to the maintenance of that degree of heat which is necessary to health and life. But, unfortunately, many of the contrivances to this end are adapted to but one object, namely, *procuring* heat. And since this object can be accomplished most easily by employing those means to create warmth which are attended with the least circulation of the air, we are frequently heated while we are poisoned.

445. To procure heat, to distribute it, to retain it, and to obviate its attendant inconveniencies by ventilation, comprise the objects necessary to a comfortable existence in cold weather. And whatever the methods adopted for the warming of rooms, two general considerations are essential to health and comfort. These are, first, to maintain the purity of the atmosphere; second, to supply the necessary warmth. To procure these conditions it is not necessary that the atmosphere should always be heated to a temperate warmth. On the contrary, the faculties of healthy persons are more active in a moderate degree of external temperature, and respiration more satisfactory, because the volume of air inhaled affords more oxygen. Generally, it is better to obtain heat, as far as practicable, by respiration and clothing rather than by a hot atmosphere. A very warm atmosphere, whether produced artificially in winter or naturally in summer, is always relaxing and debilitating.

446. Of the various means of supplying the degree of warmth necessary to health, and of the different qualities of fuel used for the purpose, there are certain conditions worthy of special attention. It is well known that the quantity of moisture contained in fuel affects the amount of heat it produces. When wet

coal or green wood are added to a fire, they gradually abstract from it a sufficient amount of heat to convert their moisture into steam before they can be burned. And as long as any part of the moisture remains in the fuel, the fire is dull and the heat feeble. The consumption of wet coal or green wood is less rapid, but to produce a given amount of heat a far larger proportion of the fuel must be consumed than if it were dry; because, not only the steam, but the different gasses evolved during combustion affect the usefulness of fuel in proportion to their quantity and according to their capacity for heat. It is, therefore, far from true economy to burn wet coal or green wood on the supposition that because they are more durable they will in the end prove cheaper. Moreover, as moisture with fuel requires a portion of the heat generated to convert it into steam, and as steam has a great capacity for heat—absorbing a great quantity which would otherwise be distributed—it is obvious that the custom of placing pans of water over furnaces is deceptive.

447. Furnaces, as usually constructed, are liable to become red-hot whenever a strong fire is required. Red-hot iron rapidly consumes oxygen, and if the air in which red-hot iron is placed be supplied with vapor, the facility with which oxygen is consumed is much increased.

448. Water is placed near furnaces *ostensibly* to supply moisture, it being assumed that the highly heated air by furnaces is deficient of natural moisture and too dry for healthy respiration. Such air is, indeed, very injurious and very irritating, but the remedy is worse than the disease. And if a furnace cannot be so constructed as not to consume oxygen or require vapor to aid it in doing so, under the pretext of supplying moist-

ure which is provided at the expense of the oxygen, it were surely better to have no furnace.

449. The amount of watery vapor in air depends upon the temperature, and experiment has determined the force of the elasticity as well as the quantity of vapor for given degrees of heat in the atmosphere; the elasticity of the air being increased in proportion to the amount of vapor present.

450. A cubic foot of the atmosphere, at the temperature of sixty degrees, contains six and a quarter grains of watery vapor, and the proportion of vapor is doubled for every twenty-one degrees of temperature.

451. Another erroneous practice in regard to furnaces as well as stoves, is that of closing the damper when smoke has ceased to be given off. This practice is highly pernicious. Burning coals, even after they have ceased to smoke, always give out carbonic acid in large quantities, and if this gas is not allowed to escape up the chimney, it contaminates the air of the apartment.

452. By applying these truths to rooms heated by closed furnaces supplied with water, we perceive that instead of oxygen which has been consumed by the red-hot iron, the air is intensely rarefied with vapor and filled with carbonic acid, both dangerous conditions to health and life.

453. The atmosphere contains within itself the means of its own purification; and, if allowed free scope, it slowly but certainly neutralizes all foreign substances. True and effectual ventilation consists in the free access of *pure* air, and whatever of the many ingenious contrivances for warming air, if they do not have this for their aim and end, they are proportionately ineffectual and worse than useless.

454. The NERVOUS SYSTEM is the highest distinction of the animal, and in man, it is that part of the organism to whose welfare every thing else is subordinate. It is through the medium of the nervous system that the involuntary muscles concerned in digestion, circulation, and respiration are stimulated to the performance of their respective functions, it is also through the same means that we feel and think, and set the voluntary muscles in action ; and it is the link to all the powers of the mind.

455. The nervous system consists of the *brain, spinal marrow*, and *nerves*, and these together form an apparatus to which there is nothing analogous in the plant or in the lowest orders of animals. It is divisible into two chief parts, a *centre* and a *periphery*. The former consists of the brain and spinal marrow, the latter of the nerves in communication with this centre ; the periphery, however, is also divisible into two parts—the centripetal, which terminate in, or receive and convey impressions *to* the centre ; and the centrifugal, which transmit impressions *from* the centre, and stimulate the powers of motion. Besides these two sets of nerves, there is intricately blended with them a third class called the SYMPATHETIC, whose office it is to combine and harmonize the muscular movements immediately concerned in the maintenance of life, and to bring these movements into relation with the MIND—which is the grand purpose of the whole system of nerves.

456. The brain, which has for its chief function the perception of impressions, is of itself insensible, and can be excited in no other way than by impressions transmitted through the nerves. If it were possible for a human being to come into the world with a brain perfectly adapted to be the instrument of mental opera-

tions, but without nerves to transmit impressions, there is every reason to believe that the mind would remain dormant, like a seed buried deep in the earth.

457. The spinal marrow presides over many of the functions essential to life, yet it, like the brain, must be connected with all the organs over which it presides by the nerves, and without this connection of the centre with the periphery of the nervous system, the functions of the body would be alike dormant with the faculties of the mind.

458. The nerves, which constitute the periphery of the system, and supply those parts of the body which, by contact with foreign substances, are susceptible to sensations, cannot perceive the impressions which they excite. For perception, the impressions have to be transmitted to the centre. Thus it is manifest that the integrity of the whole nervous system is essential to the healthy exercise of the faculties and the functions of the organism.

CHAPTER XVII.

ALIMENT AND DIGESTION.

459. In order that alimentary substances should be capable of fulfilling the purposes of nutrition, they must be transferred into definite organic combinations. Since these are composed of the simple substances of inorganic matter, it might seem possible that inorganic substances should fulfill the purpose of food for animals as well as plants. But we have seen that the germs of all organized beings spring from certain organic combinations, which are in their turn derived from a parent mass of the same kind. And, in the same way, the individual can only be sustained by an organic, and not by an inorganic food.

460. It must not be supposed, however, that inorganic substances are useless in the nutrition of animals. On the contrary, the gases and minerals introduced with the water and nutritious decoctions, together with the ashy constituents and the salts, the phosphatic compounds, lime, etc.; all conduce to the necessities of animal existence.

461. The substances required by animals for their sustenance and growth are of two kinds—the organic and the inorganic. The former of these only are commonly reckoned as necessary aliments, but that the latter are equally so is exemplified in the circumstance

of their being always associated with the former. This is the reason that the necessity for the employment of the inorganic elements in food commonly escapes notice.

462. A constant supply of aliment is needed for the support of the body after it has arrived at its full development, but in the growing state there is an additional demand, in order to supply the material for increase. This, however, does not make so much difference as at first appears; for, if the absolute addition which is made by growth to the body in any given time be compared with the amount of *change of composition* which takes place in the same period, this latter being judged of by the amount of food consumed, and the amount of the excretions which pass off by the skin, lungs, liver, kidneys, etc., it will be found to bear but a very small proportion to it.

463. During the whole period of growth there is a constant renovation of the entire organism, the life of each part being brief, in order that it may be renewed and readapted to circumstances. Hence it is that children require a larger amount of food in proportion to their bulk than adults.

464. In advanced age the functions are slow, and the waste and demand correspondingly so, consequently aged individuals require a smaller proportion of food than other persons. Generally speaking, everything else being equal, animals require food in proportion to their necessity for maintaining heat in the body, which varies according to external temperature. In this way external cold becomes a source of demand for food, and artificial warmth may be made to supply the place of deficient nourishment.

465. Any cause which creates an unusual drain or

waste in the system, creates a demand for food. From this cause a mother who has to supply food to a nursing infant, requires a proportionate increase. Some diseases act in the same manner, and require an extraordinary amount of food for the support of the system. But besides these conditions, there is a constitutional difference in individuals, the functions in some being more rapid than in others, require a more frequent as well as a more abundant supply.

466. The compounds which ordinarily constitute the food of man, may be classed under four heads :

I. The *albuminous* or *nitrogenous* group, comprising all substances from both the animal and vegetable kingdoms which are allied to albumen. They are applicable to the maintenance of organic life by fulfilling all its requisites, namely, formation, sustenance, and the production of animal heat. The elementary composition of the albuminous compounds is identical, and by digestion they are all capable of being reduced to the same condition. They consist of a large proportion of nitrogen united with oxygen, hydrogen, and carbon ; and it is a matter of little consequence except as regards the proportion of inorganic matters with which they may be united, whether they are obtained from the *casein* of milk, the *albumen* of eggs, the *flesh* of animals, the *gluten* of wheat, or the *legumen* of peas.

467. II. The *oleaginous* or *non-nitrogenous* group, including oily substances. These are also derivable from both the animal and the vegetable kingdoms. In these the hydrogen and carbon predominate, with a small proportion of oxygen and an entire absence of nitrogen. They are principally subservient to the production of animal heat.

468. III. The *saccharine* group. These are wholly

derivable from the vegetable kingdom, and are, in their composition analogous to sugar, consisting of water and carbon. To this group belong starch, gum, wood-fibre, and the cellulose of plants, all of which closely resemble each other in the proportion of their elements, and may be changed into sugar by a simple chemical process. None of the saccharine group are directly convertible into organic tissues, but they undergo a metamorphosis in the animal body by which they are changed into the same constituents as the oleaginous group, whence, like these latter, they are in part applicable to the formation of fatty and nervous tissues, and in part to the maintenance of animal heat.

469. IV. The *gelatinous* group, derivable from the animal kingdom only, and in their composition closely allied to gelatin. These are nitrogenous, but wholly different from the albuminous group. They seem to occupy a middle position in regard to the albuminous group, with which they are identical in composition; and the oleaginous group with which they are identical in use. The use of gelatin as food is never necessary, inasmuch as all that the system requires is supplied by the albumen. The nutritive qualities of gelatin in the form of soups, jellies, and the like have been very much overrated, when thus taken it is always in excess, and it is speedily eliminated by the kidneys.

470. The apportionment of the albuminous, oleaginous, and saccharine compounds, which constitute the essential elements of food, is of the first importance, and these substances are, consequently, made attainable from either the animal or vegetable kingdom, the only requisite being that they shall be applied in the needful proportion. If man be so situated in regard to vegeta-

ble productions as to be able only to obtain such as can but partially supply these compounds, the cravings of his appetite indicate the necessity of resorting to animal food in order to supply the deficiency. Thus a diet, whose staple consists of potatoes or rice, contains too small an amount of albuminous matter in proportion to the farinaceous ; but if to this a quantity of meat be added, the proportion is assimilated to that which exists in wheat bread, which is usually taken as the standard for man's alimentation in all but extremely cold climates. The failure of wheat bread to supply all that the system requires in low temperature, depends upon the increased necessity for oleaginous substances in order to maintain heat. Bread made of Indian corn will supply this deficiency.

471. While animals and vegetables are universally distributed, they are, nevertheless, in their nature, limited to particular localities. MAN only is adapted to all climates, and his migratory nature could not be sustained without a corresponding adaptation of the production of various climates to his sustenance. If otherwise, man would be like unto the lower animals, restricted to particular localities.

472. The universal distribution of food most capable of sustaining life indicates by its kind that which is most appropriate to the various circumstances to which man is by nature adapted. Each country has a certain amount of alimentary substances peculiar to it, and the best rule is to adopt the diet of the natives in the midst of whom we dwell.

473. The kind of food man lives on not only influences his physical development, but also his character and mental capacity. The experience of mankind seems to justify the conclusion that, as a general thing, per-

sons who eat meat in certain proportions are not only more vigorous and active, but that their intellects are better developed than those who are nourished exclusively on vegetable productions. But it would be more just to base this conclusion upon the proportion of nitrogenous and non-nitrogenous compounds used in food, regardless of their derivation from the animal or vegetable kingdom. Investigation shows that many vegetable substances are prolific in nitrogenous materials, and there can be no question that *these*, and not flesh, are the true requisites of man's nature.

474. The omniverous *capabilities* of man are no other criterion than of his powers, surely, not of his excellences. The history of tribes of men who subsist wholly on animal food, shows them to be wilful, brutal, and barbarous ; while there are numerous instances of the greatest intellectual capacity among men who have wholly excluded it. The arguments derived from the teeth and the digestive apparatus are in no respect available to establish anything more than man's capability of sustenance on *either* or *both* animal and vegetable diet. Though man is the summit, the very crown and triumph of the works of the Creator, and in both the excellence of his organism and the powers of his intellect combines the perfection of development, he is not on this account so constituted as to make the utmost of his capabilities a necessity to his excellence.

475. Chemical analysis of the kind and quantity of food habitually taken by persons in the highest state of health, shows that in the gross it is no richer in the necessary materials for the support of the system than many compounds wholly consisting of vegetables, and that by no possible contingency can there exist such

an abundance of the essential elements necessary to animal life in any meat diet, as is common to several of the single articles of vegetable production.

476. The relative value of different articles of food depends upon the amount of material they contain applicable to the formation of tissue and the production of animal heat. But this amount must be measured by the powers of assimilation to the wants of the system, rather than by the quantity actually present in the substance used. Thus an aliment abounding in nutritive matter may be inferior to one containing a smaller proportion, if only a part in the former, and the whole in the latter, be readily absorbed in the process of digestion.

477. The proportion of albuminous or nitrogenous matter in any substance may be regarded as a standard of its formative value ; and the proportion of oleaginous or non-nitrogenous matter, the standard of its heat-producing power. The former of these groups furnishes the whole of the elements of *proteine*, with the necessary salts, and the latter principally consist of carbon and nitrogen. The *proteine* elements are those which are essential to all organic bodies ; and inasmuch as these are of the *first* importance, the Greek word *proteine*, signifying, "I occupy the first place," has been chosen to designate them.

478. The respective value, in weight, of the nitrogenous and the non-nitrogenous compounds to the necessities of the human system is, as one of the former to seven of the latter. The mixture of these compounds, together with the inorganic matters which exist in connection with them, comprehend all the elements essential to the sustenance of man.

479. The quantity of food required for the main-

tenance of the human body in health varies so much with the age, sex, constitution, and habits of the individual, that no standard can be fixed. The appetite is the only sure guide ; but its indications should not be misinterpreted. To eat *when* we are hungry, and to drink when we are thirsty, is a natural inclination ; but to eat *as long as* we are hungry or thirsty is a very different thing. The feeling of hunger or thirst depends less upon the state of the stomach than upon the condition of the system ; hence, it is evident that the ingestion of food cannot *at once* produce the effect of allaying hunger, though it will do so after a short time. So that, if we eat too rapidly, we may continue to swallow food long after having taken as much as will really be required by the wants of the constitution, and every superfluous particle is not merely superfluous but injurious.

480. Mastication is important not only for the production of the digestive fluids supplied by the mouth, but to prolong the meal and give time for the system to experience a supply of its wants, so that the demand may be abated in due time to prevent the ingestion of more than is necessary.

481. It is not enough for the healthy support of the body that the food ingested should contain an adequate proportion of alimentary constituents ; it is all important that these should be in a wholesome condition.

482. The greatest economy of alimentary material is exercised when the substances used contain a sufficient quantity of nutrient matter to supply the wants of the system with the least labor to the functions. And as milk is the sole nutriment prepared by nature for the young mammalia, the composition of *it* seems to indi-

cate what substances are best adapted to nourish and sustain the body.

483. Milk consists of a mixture of albuminous, saccharine, and oily substances, and when reduced to its elementary constituents, these are very nearly the same as those which may be obtained by an analysis of blood. The amount secreted also corresponds to the same conditions as affect the quantity and quality of blood, both alike depending upon a sufficient quantity of wholesome food. To obtain healthy milk without wholesome food for the animal producing it is impossible, and wholly incompatible with the plainest principles of physiology.

484. In parts of our western country a disease sometimes prevails among cattle, called "*milk-sickness,*" the "*tires,*" the "*slows,*" or "*stiff-joints*"—all these names are applied to the same disease. It is probably due to some species of *fungi* taken with their food. The milk and beef of cattle thus affected are poisonous. The blood of animals dying of milk-sickness, does not coagulate, showing a corresponding condition of this fluid with the poisonous milk secreted. These analogies should teach us the necessity of wholesome food for all animals which supply us with important articles of diet. But most of all, as regards *milk.* The residue of spirit distilleries called *swill,* commonly used as food for milch-cows, contains a large percentage of acescent material similar to putrid vinegar. The milk of cows fed upon it acquires analogous properties, while it is lacking in the necessary phosphates for healthy organization. The result is, that children fed upon it ordinarily become exceedingly voracious, on account of the deficient elements of nutrition, and take large quantities. The great amount of acid ingested acts upon the

lining membrane of the stomach and bowels, softens it and predisposes to bowel complaints, cholera morbus, cholera infantum, etc. The deficiency of phosphates absolutely necessary for the nutrition of the brain and other nervous tissues, and, also, for the hardening of the bones, predisposes to dropsy upon the brain, idiocy, and rickets.

485. Omitting the consideration of all deleterious mixtures which articles used as food may have received from various external sources, alimentary substances may acquire a poisonous character from changes taking place in their own composition. "Sausage poison," "ham poison," "cheese poison," etc., are familiar examples. *Putrid* food acts in a similar manner. All such articles should be rigidly excluded, as they are often violently poisonous, frequently varying in their effects, however, according to hygienic conditions and constitutional tendencies. Their injurious qualities are most frequently experienced among individuals habitually living in badly ventilated apartments or in miasmatic districts. Under such circumstances impure food often lays the ground-work for severe epidemics.

486. Wholesome water is no less important than wholesome food. As a necessary drink, and for culinary purposes, water contributes special qualities which are incessant and inevitable. Good water may be described in general terms as that which is fresh, limpid, and without odor ; it possesses a taste characterized by freedom from all disagreeable qualities—it is neither insipid, sour, salt, nor sweet. And if it is *soft*, it dissolves soap without a sediment. If *hard*, the converse of this is the case—a sediment is deposited, which is evidence of the presence of saline matters.

487. *Soft* water is more conducive to health for the generality of persons than *hard*, because it is a better solvent of alimentary materials. But for the same reason it is more likely to hold foreign substances in solution, and the presence of a very small quantity of some minerals—lead, for example, or putrescent matters—may escape attention. The habitual use of water containing such substances for a length of time is often followed by the worst consequences. Indeed, the use of water containing putrescent matters is sometimes speedily followed by the most pernicious results—*cholera* and other epidemics often originating from this cause.

488. *Hard* waters always contain saline constituents, and for persons of weak digestive organs the continuous use of water thus impregnated is apt to induce disease.

489. The various BEVERAGES employed by man, chiefly consist of water, holding other matters in solution. A peculiar problem, indeed, to the psychologist in the circumstance that, wheresoever the human race is found—in the highest condition of civilization as in the first dawnings of culture—the custom everywhere exists of transporting themselves by various means into a high condition of mental activity, which in its excessive and evil phenomena is called *drunkenness.*

490. Nothing can be clearer to the unprejudiced observer than that the introduction of *intoxicating* agents, of whatever kind, into the system, perverts the action of the mind, disorders the usual sequence of phenomena most purely psychical, and occasions new and strange results, which are at total variance with healthy action.

491. Alcohol, on being introduced into the stomach, is immediately absorbed, and being perfectly miscible

with the blood, speedily pervades every part of the system; and this is the case, be the quantity ever so small. In its effects, alcohol has specific influence on the brain and nervous system, and it is owing to this peculiarity of its action that it produces that singular species of delirium called *drunkenness*. It constantly seeks out and fastens upon the most sensitive portion of the animal economy, and it is owing to this that the energies of the constitution are speedily roused into resisting and eliminating it. This contest of the system, with an unnatural impression which it strives to get rid of, constitutes the *stimulating* effect of alcohol. And though this excessive action of the vital powers may be kept up for a time by the repetition of the dose, it is always sooner or later followed by a proportionate degree of exhaustion, and in proportion to the frequency of the paroxysms are the powers of the constitution lessened, and the susceptibility to disease increased.

492. Alcohol is the predominant principle in all fermented liquors, but in WINE, ALE, PORTER, etc., its character is so modified by combination with other substances that it exercises a different power over the system from what it does when used in the form of distilled liquor. When thus used, the amount of alcohol taken is not only less powerful than an equivalent portion taken as distilled liquor, but it is much slower in its action, and longer in duration. This modified action of alcohol is for the most part due to the *extractive* matters of the fruits and vegetable substances held in solution by it, yet it is none the less injurious on that account. The result of its habitual use under these circumstances is excessive nutrition or *plethora.*

493. The first effects manifest in excessive nutrition display an exuberance of health, hence the mistake that

the most perfect health is compatible with the moderate indulgence in, or the habitual use of these beverages.

494. When the supply of nutritive material is habitually abundant, and the functions of the system are stimulated, the usual effect is increase of bulk, especially if the habits of exercise are not such as to create an amount of excretion proportionate to the inordinate supply of nutritive material. If, however, the excess be only slight or casual, with a proportionate degree of physical exertion, the self-adjusting powers of nature may be equal to the irregularity, and prevent the transition of healthy into diseased action. But if the excess be great or habitual, the organic functions are overtaxed, and their conservative powers necessarily languish. This condition is succeeded by such irregularities as display the worst effects of the beverage, by the production of incurable diseases, which have had their foundation in apparent "perfect health."

495. It is obvious, therefore, that the exuberance of health evinced by the florid countenance and fatness of persons addicted to the habitual use of fermented liquors, are the suspicious evidences of a constitution taxed to the very highest degree of forbearance, which must, in course of time, become relaxed, and sink even below the normal standard of resistance. In this vitiated state of the system, constitutional predispositions to disease, that might otherwise have lain dormant, are frequently roused into speedy fatality.

496. That there are some constitutions which appear to be unaffected by the habitual use of alcoholic liquors is no less true than that of any other habit *tolerated* by the natural powers of endurance. But such habits only serve to demonstrate the capabilities of the human con-

stitution, and are in no event admissible evidence of natural adaptation.

497. The composition of alcohol most nearly approaches that of the oleaginous group of alimentary compounds, and it may, therefore, be considered as possessing heat-producing qualities. But in this regard it should be born in mind that, while alcohol is heat-producing, this quality chiefly consists in its own combustibility, or, in different words, in its quicker miscibility with and circulation in the blood than any other heat-producing substance. In virtue of this, carbonic acid and other injurious substances are retained until the alcohol is consumed or passes off. The blood, therefore, loses its usual facility of decarbonization, and retains a dark venous aspect by the retention of carbon. Hence there can be no justification in the use of alcohol to maintain animal heat, unless there is a deficient supply from such other substances as will not hinder the elimination of carbon, the undue retention of which is always injurious. Such circumstances, however, do sometimes arise. Dr. Kane informs us that when *short* of *oleaginous food*, in excessive low temperature, a small quantity of brandy carefully served out in *spoonful doses*, was invaluable. This is the experience of others in similar emergencies, namely, *when unable to obtain food under excessive fatigue and in severe cold*, alcohol becomes valuable as a *temporary* heat-producing agent, but as a reliance, or in continued exertion, it does harm by the consecutive depression.

498. In some persons there is a fixed constitutional debility, on account of the early habitual use of alcohol, which apparently deprives it of its usually stimulating qualities. In such persons, the continued use

seems to be practiced with more impunity, and if it is suddenly left off serious results sometimes follow. When such persons are, by prison discipline or otherwise, denied an abuse which has to them become a necessity, their vitiated constitutions incapable of sustaining any hardship, speedily sink unless stimulated by alcohol. But these cases become the care of the physician and the supply of *medicine* the sphere of the apothecary.

499. TEA, COFFEE, AND MATÉ.—These three substances, obtained from plants of totally different aspect, present a relation to each other no less singular than their immemorial and universal use. To inquire into the origin of their use, would be an interesting though perhaps tedious topic. The Chinese have a legend which runs thus :—"A pious hermit, who in his watchings and prayers had often been overtaken by sleep, so that his eyelids closed in holy wrath against the weakness of the flesh, cut them off and threw them on the ground. But a god caused a Tea-shrub to spring out of them, the leaves of which exhibited the form of an eyelid bordered with lashes, and possessed the gift of hindering sleep." It will suffice to state that the use of tea as a beverage in China can be traced back to the third century, and that when the Europeans first became acquainted with it, it was in common use all over the South of Asia.

500. Coffee, also, was first used, according to the tradition of the Abyssinians, by those who wished to keep themselves awake during the holy nights of prayer. It was originally drawn from a large earthen vessel into small cups, which were handed round during religious services, and on this account the enmity of the Mohammedans was excited against its use. And

they went so far as to affirm that the countenances of those who drank coffee would, on the day of resurrection, appear blacker than the coffee grounds. Coffee was introduced into Arabia from Abyssinia about the beginning of the fifteenth century.

501. Maté, or Paragua-tea, the leaves of a Brazilian holly *(Ilex Paraguayensis)*, has been the customary drink of the Brazilians from time immemorial. Its properties are similar to those of tea and coffee.

502. Thus have all these beverages become everywhere necessaries of life, and everywhere is the origin of their use enveloped in obscurity. Everywhere has man, not led by rational considerations, by knowledge of their properties and action or by comparison of them with known nutritive substances, but as it were instinctively, added them to the number of his daily wants. The strangeness of this circumstance has not failed to attract the attention of physiologists and chemists, and the results of their inquiries have been no less astonishing than singular. Tea, coffee, and maté are all found to contain a crystallizable substance called *theine* or *caffeine*, remarkable for its identity. This substance chiefly consists of nitrogen, and when used alone, is devoid of any striking action on the system. Yet experiment and observation justify the conclusion that the use of tea, coffee, or maté serves an important purpose in the organism by *retarding the waste of the system, and in such wise limiting the demand for food as to make a limited amount go further.*

503. Cocoa, chocolate, chicory, etc., all yield wholesome and nutritious beverages, consisting of various proportions of oleaginous and extractive matters, but they are deficient in the peculiar properties of tea and coffee.

504. The satisfying effects of TOBACCO in individuals long addicted to its use, are in many respects analogous to the effects of tea, coffee, and maté. At least, sailors and soldiers are found to sustain the privation of food much better when they are supplied with tobacco.

505. There is nothing which better exemplifies the *effects of habit* than the use of tobacco. On once becoming fully addicted to its use, tobacco becomes such an absolute necessity that the greatest distress arises from its privation. In the wintery solitudes of the Laplander or the Esquimaux, it appears to be essential to their existence. The Arab in the desert would as soon part with his *gourd* as his pipe. And the sailor, grant him but this one luxury, and he will defy the fury of the elements. Nor is tobacco less worshiped at the shrine of fashion. The fascinating influence of its effects once thoroughly incorporated, under whatever circumstances, is so wedded to the functions, that health may not only suffer by its deprivation, but even life itself become, at least, a burden. The following case is related by M. Mérat, in the *Dictionaire des Sciences Medicales.* "I recollect about twenty years ago, while gathering samples in the forest of Fontainbleau, I met a man stretched on the ground, who seemed to me to be dead, but, on approaching him, he asked if I had any snuff; on my replying in the negative, he sank back almost in a state of insensibility. He remained in this state until I brought a person to him, who gave him several pinches, after which he informed us that he had set out on his journey that morning, supposing that he had his snuff-box along with him, but he soon found he had left it behind; that he had traveled as long as he was able, till at length, overcome by distress, he found

it impossible to proceed farther, and without my timely aid he would certainly have perished."

506. When the use of tobacco is carried to excess, general lassitude and indisposition to any exertion, either mental or physical, followed by tremor and excessive perspiration, are the usual effects ; and if the abuse is persisted in, more serious evidences of injured health succeed.

507. An erroneous notion prevails that smoking or other use of tobacco will prevent the poisonous influence of malaria ; so far from this, the excessive use of tobacco, or the use of it by persons unaccustomed to it under such circumstances, renders the system even *more* susceptible. The poison of malaria is usually potent in proportion to whatever departure from the most perfect standard of health and the powers of endurance, and inasmuch as tobacco can in no respect heighten these powers, except when persons have been long addicted to it, there is no reason to suppose that it can be of benefit.

508. In all cases, the *first* effects of tobacco impair the tone of the functions, and it is at least doubtful whether a full development of the most healthy exercise of all the functions can ever take place when subject to its influence. And when we consider the many disagreeable concomitants, independently of the injurious effects, we see in them ample reason to discountenance its use.

509. It is related of Dr. Franklin, that a short time before his death, he declared to one of his friends that he had never used tobacco in the course of his long life, and that he was disposed to think that there was not much advantage to be derived from it, for he had never known a man that used it who advised him to follow his example.

510. The sources of food for animals are so far different from those which obtain in plants that a preparatory process becomes necessary for the reduction of alimentary material to a fluid form. This process is effected in the cavities of the body, which are bounded by a continuation of its external surface, modified by the power of producing certain solvent liquids for the solution of aliment, and also by an elective faculty for the selection of such parts as are best adapted to the nutrition of the fabric.

511. So long as aliment remains unabsorbed it cannot be regarded as introduced into the system, since it merely holds the same relation to the absorbent vessels as the nutritious fluid in which the roots of plants may be immersed, bears to the ducts which they inclose ; in either case, the material has to be *introduced into the circulation* before its effects are manifest.

512. The relation between mere presence and absorption may be elucidated by the elective rejection of certain poisons. The poison of the most venomous serpents, and the *woorari* of the South American Indians, when taken into the stomach, are perfectly harmless ; the lining membrane of the alimentary canal showing a peculiar inaptitude for allowing them to enter the absorbent system ; yet the most minute quantity of either of these poisons when introduced into the circulation, even by the slightest scratch, produces certain sudden death. Most other poisons are absorbed with great facility, and on this their potency chiefly depends. Fluidity, however, is the first necessity of absorption, and the ample provision with regard to this for the purposes of nutrition, constitutes the function of digestion

513. Digestion is the function of reducing the nutritive material to a fluid state. This function is accomplished by an organic apparatus, composed of a series

of actions. Enlargement of the free and active surface constitutes one of the most striking advantages of subdivision; a pulverized body being much more easily dissolved than a large piece of the same substance, because more molecules of the solid and the fluid are brought into immediate contact. When food is received into the mouth, saliva is immediately poured out on all sides, and if it is of a substance easy of solution, it is at once reduced to a liquid; if not, however, it is masticated or minutely divided by the teeth, and thoroughly incorporated with the solvent fluids of the mouth, after which it is swallowed. In the stomach the food is exposed to still more powerful solvents, and is there converted into a homogeneous pulp, termed *chyme.* As fast as the chyme is formed it escapes into the intestine, where a still further digestion occurs through the admixture of yet more powerful solvents—the *bile* and *pancreatic* juice; these effect perfect solution, and convert the alimentary material into *chyle,* a milky substance that admits of absorption. The chyle is first absorbed by the lining membrane of the intestine, and from this arise numberless little vessels which convey it to a large duct, which connects with and empties into a vein near the heart. The chyle being the nutritive portion of the food is thus conveyed into the circulatory apparatus to be converted into blood, and applied to the purposes of nutrition.

514. Blood is a fluid of definite constitution, chiefly formed of chyle. When examined under the microscope, it is found to consist of two distinct parts; one of a yellowish transparent liquid, chiefly composed of water, named *serum;* and another of a number of little round cells, called *globules*—these swim in the serum. Blood, then, in its ordinary state, is a watery fluid,

holding solid globules in suspension. Chemistry teaches us that the blood contains the greater part of the substances that enter into the composition of the different tissues of the body. Wherefore the blood may reasonably be regarded as containing all the material necessary for the formation and nutrition of the body, and as well meriting the designation of *flowing flesh.*

515. The blood not only distributes all the material necessary for nutrition, but it is by means of the blood, also, that the injurious or non-essential particles of the substances taken with food are thrown out. To accomplish these purposes, the blood must unceasingly traverse all the organs and penetrate all the tissues, communicating nutritive properties and receiving in exchange for them impurities, and dispose of these impurities before performing another circuit.

CHAPTER XVIII.

CIRCULATION OF THE BLOOD.

516. In the circulation of the blood, the heart forms the centre or chief agent. It drives the blood in the peripheric or centrifugal direction within special tubular conduits, the arteries, and receives the blood back again by the veins, within which the fluid returns with a centripetal course.

517. The blood on being returned to the heart, is poured into the right auricle or receiving cavity. From this it passes into the right ventricle or propelling cavity, thence through the pulmonary arteries into the lungs, where it is freely exposed to the air, and then through the pulmonary veins it is conveyed into the *left* receiving cavity or auricle. From the left auricle it passes into the left ventricle, this powerfully contracts and forces the blood into the great systemic artery, the *aorta*. The aorta gives off numerous branches that subdivide and ramify all the tissues of the body, and ultimately terminate in small hair-like vessels, termed *capillaries*, these connect the arteries with the veins. It is through the capillaries that the blood comes into immediate relations with and nourishes all the tissues, and it is also through them that impurities are eliminated, and the blood purified.

518. It is obvious from this description that the two

sides of the heart contain different kinds of blood. The *right* side contains dark-colored or *venous* blood, the *left* red, or *arterial* blood. The right ventricle, the pulmonary arteries, the capillaries of the lungs, the pulmonary veins, and the left auricle, all belong to the respiratory or *pulmonic* circulation, while the left ventricle, the aorta, the arteries of the body, the capillaries, and the veins of the body, and the right auricle, belong to the great or *systemic* circulation. And since the right side of the heart in the human adult is completely shut off from the left, the blood must pass along each of these paths successively, before returning anew to the same part of the heart.

FIG. 42.

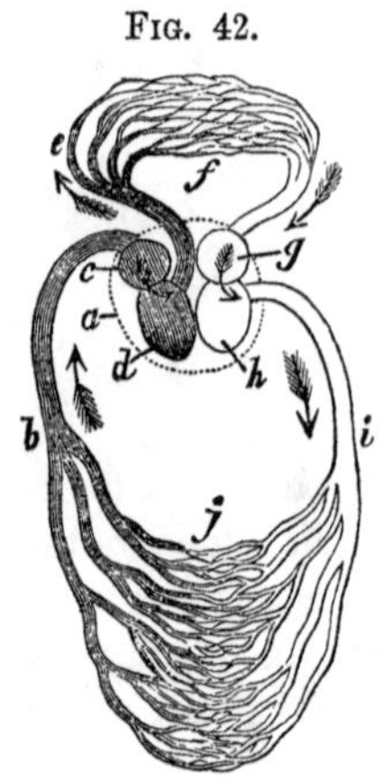

DIAGRAM OF THE CIRCULATING APPARATUS IN BIRDS AND MAMMALS.

a, the heart containing four cavities; *b*, vena cava, delivering venous blood into *c*, the right auricle; *d*, the right ventricle propelling venous blood through *e*, the pulmonary artery to *f*, the capillaries of the lungs; *g*, the left auricle receiving the ærated blood from the pulmonary vein, and delivering it to the left ventricle *h*, which propels it through the aorta *i*, to the systematic capillaries *j*; whence it is collected by the veins and carried back to the heart through the vena cava.

519. The heart forms a forcing and sucking pump, which drives its contents onwards through the tubes of the vessels. By its strong muscular walls it is endowed with a capacity of alternate contraction and extension. Its contraction furnishes the force of pressure which impels the contained blood, while its relaxation affords space for the entry of a new supply. The two states of contraction, or *systole*, and relaxation, or *diastole*, continually alternate with each other, and a

beat of the heart comprises that space of time in which each of the four cavities has undergone one systole and one diastole.

520. The quantity of blood contained in the human system may be set down at about one-thirteenth of the weight of the body, but the amount varies with the mode of living and the habits of the individual. Indications of excess are marked rather by an exuberance of health than of disease. The functions of the body act more vigorously for the purpose of expelling the excess, which the ordinary powers are unable to appropriate. An excess of blood is generally the result of supernutrition and deficient exercise; and by its long continuance from these causes, health and life may be forfeited. A deficiency of blood is indicated by palor, languor, and emaciation.

521. The quantity of blood contained in the heart at any one time probably never exceeds ten ounces. It has been ascertained by experiment that each beat of the human heart impels about $4\frac{2}{5}$ ounces of blood into the systemic, and the same quantity into the pulmonic circulation. Estimating the average number of beats at seventy per minute, the quantity of blood propelled in this unit of time may be stated at three hundred and eight ounces, showing that the whole mass traverses the entire body about once a minute.

522. The wonderful activity displayed in this function may be better appreciated by estimates proportioned to greater lengths of time. It will then appear that the heart contracts more than four thousand times an hour, and that as each contraction sends forward $4\frac{2}{5}$ ounces of pure blood, over one thousand pounds of this fluid pass through the heart every hour!

523. When the blood has completed one tour of the

system, it necessarily passes through the lungs before beginning another. This route is intimately connected with the purification of the blood. When the blood leaves the heart by the systemic arteries, it is arterial or *pure* blood and of a *bright* red color, but when it returns by the systemic veins as venous blood, it is loaded with impurities which have changed its color to a *dark* red.

524. The change from arterial or pure blood is effected in the capillaries of the system; in these the blood gives off the necessary constituents for the nutrition of the body, and receives in exchange for them the effete or worn out material which is no longer of use, and which, if retained, is thenceforth poisonous.

525. Loaded with impurities, the blood is returned by the veins to the right side of the heart, where the pulmonary artery receives it and conducts it to the capillaries of the lungs. In these it is freely exposed to the air, the impurities given off, oxygen absorbed, and the blood again changed to a bright red color. The pulmonary veins then conduct this renovated blood to the left side of the heart whence it again sets out for a new circuit.

526. One of the chief products of the decay of tissues consequent upon the loss of vitality is carbonic acid. This is set free during the decomposition of every kind of organic matter, the carbon of the substance combining with the oxygen of the atmosphere. Carbonic acid is also generated from the carbon contained in food. This reaction between living beings and the air may be included under the general term of *œration*, which is chiefly accomplished through the medium of the fluids of the body by the respiratory apparatus.

CHAPTER XIX.

RESPIRATION.

527. THE FUNCTION OF RESPIRATION essentially consists in the evolution of carbonic acid from the blood, and the absorption of oxygen from the surrounding medium.

528. The general principle of the function of respiration is this : the lungs are suspended in a closed cavity bounded above and around by a bony framework—the ribs—the spaces between being filled up by muscles and membranes, and below, by the *diaphragm*, a broad muscle stretched transversely between the thorax or chest, and the abdomen, which it separates from each other.

529. When the diaphragm is in a state of rest, it forms an arch with its convexity directed towards the thorax. When it is drawn downwards by means of muscular contractions, the lungs dilate by the expansion of the air contained in them, and the heavier, or outside air is drawn in or inhaled. Expiration is accomplished by the elasticity of the tissues. The relaxed diaphragm being antagonized by the pressure of the organs underneath, returns of its own accord to its previous situation. The organs in contact with the under surface of the diaphragm, playing in close contact during the whole of its motions.

530. Under ordinary circumstances, the lungs com-

pletely fill the thorax, though the capacity of this cavity is susceptible of being greatly altered by the movements of its boundaries. The mechanism of the respiratory apparatus is so arranged that the air is *drawn* into the lungs instead of being forced into them.

531. If the capacity of the thoracic cavity be enlarged from *c c* to *e e*, the fluid contained in *cd*, *cd*, adapts itself to its new position in *ef*, *ef*. A quantity of air corresponding to this enlargement will be drawn in through the windpipe. Upon this process the act of inspiration depends, while the act of expiration consists in the return of the thoracic walls from *f f* to *d d* and the expulsion of air by the windpipe.

FIG. 43.

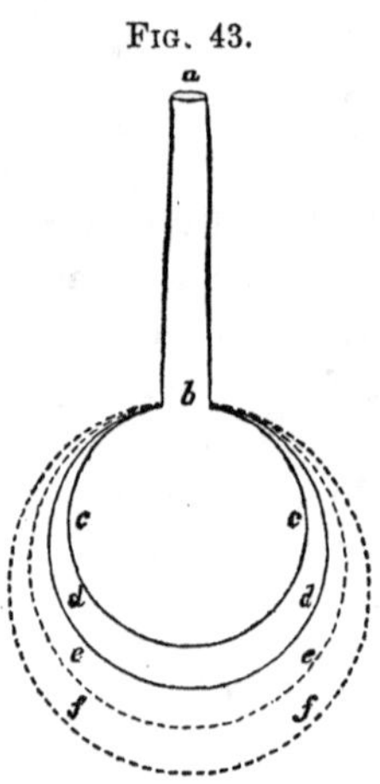

This diagram, we will suppose, represents two elastic bags supplied by a tube *a b*, which we will call the *windpipe*. The first bag, or inner line of the circle, marked *c c*, the *lungs*, and the outer bag, *d d*, the *thorax*. The space between these bags is filled by a fluid secreted by the *pleura* or membrane covering the lungs *c c*, and lining the thorax *d d*; the membrane beginning and ending at the root of the lungs *b*, where they are connected to the thorax by means of the reflected pleura.

532. By the arrangements that are made for the constant renewal of the air which the lungs contain, the respiratory surface is brought into the most advantageous use possible. The arrangement and number of the air-cells in the lungs are adapted to the greatest space, insomuch that if it were possible to convert that space into one extent of surface it would many times exceed that of the entire surface of the body.

533. The partition between these air-cells, the diameter of which varies from $\frac{1}{200}$ to $\frac{1}{70}$ of an inch, are formed by folds of membrane, between each of which is arranged a layer of capillary blood-vessels. So that the blood in the capillaries is exposed to the air contained in the cells on all sides. It has been estimated that the number of air-cells grouped round the termination of each bronchial tube in the bird is not less than 18,000, and that the total number of air-cells in the human lungs is not less than *six hundred millions*. Some idea may be formed from this estimate of the vast extent of surface provided for bringing the blood into relation with the air.

Fig. 44.

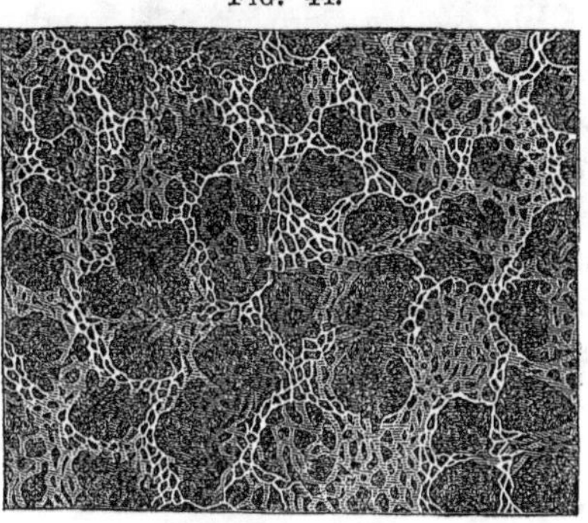

The Human Lung.

Arrangement of the capillaries of the air-cells of the human lung.

534. After respiration once takes place the lungs are never again destitute of air, and frequent attempts have been made to estimate the quantity remaining in them. It may be safely stated that one hundred and twenty-five cubic inches of air may be forcibly expelled from the lungs, while over one hundred cubic inches will still be retained, and that about forty cubic inches are added at each inspiration giving over *three hundred cubic inches as the average capacity of the human lungs.*

535. By respiration the air within the lungs is altern-

ately augmented and diminished, and by the mutual diffusion of gases, the whole quantity is renewed every time we respire pure air. Through the lining membrane of the air-cells, the air communicates with the blood contained in the capillaries, and by this means about half a cubic inch of air is absorbed every time we breathe. This half a cubic inch of air that disappears is wholly oxygen, and it is replaced by a relative proportion of carbonic acid, exhaled through the same membrane at the same time that the oxygen is inhaled.

536. The quantity of oxygen consumed in respiration varies according to the condition of the system. It may be increased by muscular exertion to fourfold what it is in a state of repose. If the exertion is excessive, however, so as to induce fatigue, the consumption of oxygen is diminished. The quantity of oxygen consumed is also affected by temperature and diet. More is required in a low than in a high temperature, and in persons who live on animal food than those who subsist on vegetables. Individuals addicted to the use of alcoholic stimulants consume less oxygen than others.

537. The average amount of oxygen consumed by healthy persons in a state of repose, being half a cubic inch every respiration, amounts in a day to about fifty thousand cubic inches or more than twenty-five cubic feet. Now, as oxygen constitutes but one fifth of the volume of the air, a single individual renders not less than one hundred and twenty-five cubic feet of air unfit for respiration every twenty-four hours.

538. The unfitness of air for respiration, however, does not wholly depend upon the abstraction of oxygen. It has gained a large accession of carbonic acid and a quantity of vapor. The amount of carbon thrown off by respiration, like the consumption of oxygen, varies

according to circumstances. It is doubled in the same individual in passing from a temperature of ninety-five to sixty-five degrees ; and in the same ratio as temperature diminishes. During active exercise and after full meals, the quantity of carbonic acid exhaled is much increased. On the other hand, it is diminished by repose and sleep. The quantity also differs according to peculiar condition of age and sex. By numerous experiments, the average quantity of carbon daily exhaled by an adult individual may be set down at *ten* ounces through the lungs, and *one* ounce by cutaneous transpiration ; making *eleven* ounces. This converted into carbonic acid, would produce about *twenty cubic feet* daily.

539. By computing the whole quantity of air respired in twenty-four hours, it is found to be about *four hundred cubic feet;* this contains five and a half per cent. of carbonic acid. This is of great practical importance in relation to the capacity of school-rooms, churches, lecture-rooms, court-rooms, prisons, etc., and above all to *tenement houses.* There should, in all cases, be *twice* as much, or at least 800 cubic feet of, space, as the *minimum* that can be safely assigned to every occupant.

540. The *exhalation of vapor* in respiration is also influenced by the physical condition of the individual. As a *rule,* expired air contains as much moisture as would be necessary to saturate it at the temperature of the body : consequently, it will vary in the inverse proportion to the quantity of moisture the air contains before it is respired. One thousand parts of vapor exhaled by the lungs contain of pure water, 907 parts ; carbonic acid, 90 parts ; and animal matter, 3 parts. It is well known that other substances, when they have

been introduced into the circulation, may be thrown off from the system through the lungs.

541. In ordinary respiration the force with which we expire tends to keep the respired air for a time in suspension. But for this force the carbonic acid would at once *fall* from its greater specific gravity than atmospheric air. The aqueous vapor suspends the animal matter; *pure* air separates and mixes in the atmosphere; the water condenses and *falls* or disperses in vapor according to atmospheric temperature, and the animal matters continue to float, fall, or decompose according to the condition of the surrounding medium.

542. In crowded and illy ventilated or artificially lighted apartments this disposition of respired air is retarded. Carbonic acid emitted by burning gas or candles is not more than half as heavy as the air, and this readily *ascends* and escapes, if unrestrained; but if restrained, it soon cools and becomes heavier, and then mixes with the other heavy exhalations, and occupies the lower strata.

543. Carbonic acid from the lungs at the temperature of the body, is one third heavier than the air at the temperature of forty degrees. And carbonic acid is lighter than the air, only when it is two hundred and fifty degrees hotter, as when it is escaping from burning gas.

544. It is scarcely necessary to dwell upon the fact that, by the repeated passage of the same air through the lungs, it may, though originally wholesome, be so strongly impregnated with carbonic acid, and may lose so much oxygen, as to be rendered utterly unfit for the maintenance of life. The barbarous act of the half-civilized nabob of Calcutta, who more than a century ago inclosed 146 persons in a dungeon eighteen feet

square, of whom 123 died in one night, has been so often printed, that every one is informed on the poisonous nature of air repeatedly respired. Yet such catastrophes, though perhaps of less magnitude, are far from being unusual, even in civilized communities.

545. In 1848, the deck passengers of the emigrant steamship Londonderry, on her passage by the way of Liverpool to America, were ordered below by the captain on account of stormy weather, and the hatches of a small cabin closed upon them. Of 150 individuals thus shut in, 70 suffocated before morning.

546. In 1858, a "distressed American seaman," late second mate of the ship Waverly, of Boston, entered a complaint at the United States Consul's office, at Seville, for protection from punishment for aiding in confining over 400 coolies below decks for *twelve* hours, during which time 270 died. This mate, a chief actor in such wholesale murder, complained of "hard treatment" because he had been sentenced to jail four years for the offense, and because the consul would not intercede to prevent it!

547. Other examples might be cited to show the loss of life from defective æration or smothering. But they fail to attract the sympathy of the multitude, unless committed by the agency of an exploded boiler or the physical power of a fiend. A nightly brawl which results in the death of a single individual, will scour the country and the seas for the murderer, while the owners and abettors of suffocating machines, known as factories, tenement-houses, or ships, are unnoticed and unknown.

548. Pure air is the great disinfecting agent in nature, tending constantly to dilute and remove all pernicious emanations from whatever source, and in

proportion as we confine or restrain it, do we generate and foster disease.

549. The processes of respiration and combustion are perpetually tending to destroy those nicely adjusted proportions of respirable air, by the abstraction of oxygen and the substitution of carbonic acid. But the natural tendency of gases to rapidly permeate each others texture, and become equally diffused is so great that the poisonous air is not allowed to accumulate, but rapidly diffuses itself; while the vital air rushes, by counter tendency, to supply whatever deficiencies take place. Hence the universal constancy of the atmospheric elements.

550. The atmosphere is subject to a law which characterizes all elastic fluids, namely, it presses equally on all sides; and when any portion becomes lighter than the others the denser portions rise into their place and force them to seek a still rarer medium, always creating a current from the point of greatest to that of least pressure. When the disturbing cause is local, transient, and irregular, partial derangement ensues on account of the action necessary to cause a speedy adjustment But as the local disturbances are always as much wanting on one side as they are in excess on the other, they are equivalent to undulations of the same medium. Their balance will still maintain an equality of pressure, which is one of the first conditions of the atmosphere. And, whatever may be the disturbing causes, the restoration of equilibrium is the object of all the motions excited.

551. In the free atmosphere its own weight is a compensating force, consequently its weight and elasticity both diminish in ascending from the surface of the earth. The rarefied atmosphere of high mountains

makes respiration laborious, and renders such localities untenable to persons of weak lungs; while mountain slopes below the condensed vapors, especially their eastern and southern sides, are of all places the most salubrious. The salubrity of such localities is due to the constant winds which incessantly renew and revivify the air. The impression of moderate cold and light winds tends to fortify and render the body agile. But *still* air, by the invisible particles of organic matter that float in it, speedily becomes like stagnant water, full of self-multiplying poison deadly to human beings.

552. The winds are a fertile and immediate source of health, because they serve to equalize temperature, and to scatter pernicious effluvia and condensed vapors, which would otherwise accumulate in particular places. But their immediate effects are favorable or unfavorable to health according to the region from which they set out or pass over.

553. The Simoon of Africa and the Sirocco of Italy owe their characters to the condition of the localities over which they pass. It is thus that while the winds bear along, and, to a certain extent, favor the development of disease beyond the place which generates it, that their benefit is most manifest. For miasm once "scattered to the winds" is in a gulf of destruction, only as complete as that which attends human beings in windless abodes.

554. The northeast winds of the Atlantic States are damp and unhealthy, because they sweep down the fogs and mists of a northern sea-shore. And the southeast and southerly winds are clear, dry, and wholesome, because they are from warmer latitudes, and from the pure surface of the ocean. *Sea air* is, indeed,

the type of purity, and the most promotive of health. In sea air, pure water is the only isolation. The deposit of salt sometimes observed on board ship is from the spray on the sails, from which it has been blown or shook. And this salt may possibly be borne up by strong winds, and thus subjected to respiration. Yet, considering the perfect solubility of these molecules of salt in the fluids of the system, and the very small amount which can, under any circumstances, be thus conveyed, it can under no state of affairs be incompatible with the most delicate organic function.

555. The salts of the sea do not communicate their properties to the air, because they are not volatile, and never exist in a state of vapor. And, owing to the salts which the sea holds in solution, evaporation of sea water takes place with more difficulty than fresh water, or of water containing substances easily decomposed by contact with the sun's rays; consequently the air is damper on sea-coasts and river-banks, and in many interior places, than it is at sea. Cold and heat are much less intense or oppressive in the same latitudes at sea, than on land; and there is a much greater equality of temperature for night and day. While from the greater purity of the atmosphere and its lowest strata, the air is more condensed and concentrated: all combining to promote the most healthy exercise of the respiratory function.

556. Light, too, is a collateral benefit of sea air; and doubtless not a little of the rigidity of tissue and hardiness which characterizes the sailor is owing to this cause. Free access of light favors nutrition and regularity of development, and contributes to beautify the countenance. While deficiency of light, especially in early life, is usually characterized by ugliness, rickets,

and deformity; and is a most fruitful source of scrofula.

557. Sea air proper possesses no deleterious qualities whatever. The ill effects sometimes ascribed to it are usually owing to the want of ship ventilation and due regard to cleanliness. Indeed, no truer maxim can be promulgated than that the want of purity is the want of health. *Malaria*, or bad air, is always an extraneous production, and its noxious effects arise from being in such a state as to be readily removable by a sufficient quantity of pure air, which, were it fully and constantly accessible, would altogether prevent any such noxious emanations.

CHAPTER XX.

FUNCTIONS OF THE SENSES.

558. The origin of all mental activity consists in receiving impressions, and being conscious of them. The impressions thus realized constitute *sensations*.

559. Sensations may be defined to be consciousness of impressions; and they may be divided into two classes: internal and external. Internal sensations are those which indicate the natural wants of the organism, such as hunger and thirst. External sensations are impressions derived from the exercise of the organs of the senses.

560. The different modifications of the faculty of sensation constitute the *five senses*, by means of which we acquire all the ideas we have of surrounding objects.

561. The five senses all have some properties in common. They are all situated on the surface, so as to be able to act with facility on external bodies. They all consist of two parts: the one physical, which modifies the action of the body that causes the impression; the other nervous, or vital, which receives the impression and conveys it to the brain. In the eye and ear we have better illustrations of this similarity than in the other senses. The physical portion of the eye is a true optical instrument, which modifies the light before it impinges upon the nervous portion or retina. A similar

modification is produced by the physical portion of the ear on the vibrations of sound, before they reach the auditory nerve.

562. In the other senses, the physical portion forms part of the integuments in which the nervous portion is situated, and cannot, therefore, be easily distinguished. Some of them again are symmetrical; that is, composed of two similar parts united by a median line, as the skin, tongue, and nose; while the eye and the ear are in pairs, and on this account better suited to the appreciation of distant objects.

563. Each organ of sense is specially adapted to a definite cycle of influences, or as it is more commonly expressed, to a suitable stimulus. To other impressions it responds either not at all, or but slightly. Thus the eye only responds to light, and not to the more forcible vibrations of ponderable matter, which give rise to the specific sensations of sound.

564. The sense of smelling necessarily presupposes the gaseous or vaporous condition; and if an odorous liquid is brought in *contact* with the membrane of the nose, the impression is then received by the sense of touch and not of smell.

565. The impulse of ponderable substances, the movement of the waves of light, the action of the odorous and sapid substances, and the change of temperature and pressure, are the chief stimuli recognized by the several organs of our senses.

566. It is a common observation, that the loss of one sense occasions greater vividness in the others. This is only true as regards the senses which administer chiefly to the intellect—those of touch, hearing, and sight. The senses of smell and taste may be destroyed, while the others will remain uninfluenced by the loss.

When there is a congenital deficiency of one or more of the senses, the mind is utterly incapable of forming any definite ideas in regard to the properties of objects only within the scope of the wanted senses. Thus an individual born blind can form no conception of color; nor the congenitally deaf, of musical tones. And in those lamentable cases in which the sense of touch is the only one through which ideas can be called forth, the mental operations necessarily remain of the simplest and most limited character, unless the utmost attention be given to the development of the intellectual faculties, and the cultivation of the moral feelings. The extent to which this *may* be accomplished, is exemplified in the case of Laura Bridgman. But a person born *deaf, dumb,* and *blind* is supposed to be incapable of any understanding; as wanting all those senses which furnish the mind with ideas, and is regarded by the law as in the same state as an idiot.

567. All the senses may be exercised *passively* or *actively.* The acuteness with which particular sensations are experienced, is influenced in a remarkable degree by the *attention* they receive from the mind. By directing the attention, we can render the mind much more active; and hence the difference between simply seeing or passive vision and attentively looking, between hearing and listening, smelling and snuffing, touching and feeling. Yet, to preserve the senses entire in the vigor and delicacy they are capable of acquiring, the impressions should not be too constantly or too strongly made. The occasional use of the sense of smell, under the guidance of the will, may be the test on which the chemist or the perfumer relies in the discrimination of numerous odorous characteristics; but if the sense be constantly or too frequently stimulated

by excitants of this or any other kind, dependence can no longer be placed upon this means of discrimination. The maxim, that "habit blunts feeling," is true only in such cases as these. *Education* judiciously applied, can indeed render the senses exceedingly acute.

568. Volition also enables us to deaden the force of sensations. By corrugating the eyebrows and half closing the lids, we can diminish the quantity of light when too powerful. We can breathe through the mouth when we wish to avoid disagreeable odors, or we can completely close the passage by the nostrils with the fingers. Over the hearing we have less command; whenever we wish to diminish the intensity of sound, the hands are always called into action.

569. It is a general rule with regard to all the sensations, that their intensity is much affected by habit, being greatly diminished by frequent and continual repetition. But this is not the case with those sensations to which attention is particularly directed, these lose none of their acuteness by frequent repetition, on the contrary, they become more readily cognizable. A good example in illustration of this is the effect of sound on sleeping persons. The general rule is that sensations *not attended to*, are blunted by frequent repetition.

570. Feelings of pain or pleasure are connected with particular sensations intensified; a pleasurable sensation, however, may be changed to an extremely painful one if violently exercised, or if it be too long continued. This is the case alike with those special sensations communicated through the organs of sight, hearing, smell, and taste, as with those transmitted by the nerves of common sensation, and there can be no doubt that the purpose of the association of painful feelings with vio-

lent excitement, is to stimulate the individual to remove from what would be injurious in its effects upon the system. The pain resulting from violent pressure on the surface, or from the proximity of a heated body, gives warning of danger, and excites mental operations destined to remove the part from the influence of the injurious cause ; while fatal accidents frequently result from the loss of sensibility, the individual not receiving the customary intimation of danger. The temporary insensibility induced by fainting sometimes leads to grave accidents from the incautious use of pungent vapors and "salts" applied to the nostrils, the individual not receiving that notice which would have prevented the injury in an active state of the senses.

571. But the most extraordinary case of injury on account of temporary insensibility, on record, is that of a poor drover recorded in the "Journal of a Naturalist."

572. "A traveling man, one winter's evening, laid himself down upon the platform of a lime-kiln, placing his feet, probably numbed with cold, upon the heap of stones newly put on to burn through the night. In this situation sleep overcame him, the fire gradually rising and increasing until it ignited the stones upon which his feet were placed. Lulled by the warmth, the man slept on, the fire increased until it burned one foot and part of one leg above the ankle entirely off, consuming that part so effectually, that a cinder-like fragment was alone remaining, and still he slept on, and in this state was found by the kiln-man in the morning. Insensible to any pain and ignorant of his misfortune he attempted to rise and pursue his journey, but missing his shoe, requested to have it found, and when he was raised, putting his burned limb to the ground to support his

body, the extremity of his leg-bones crumbled into fragments, having been burned into lime. Still he expressed no sense of pain, and probably experienced none, from the gradual operation of the fire and his own torpidity during the hours his foot was being consumed. This poor drover survived his misfortune in the hospital about a fortnight, the fire having extended to other parts of his body and rendered his case fatal."

573. TOUCH is the general feeling of sensibility possessed by the skin. But it comprehends a modification of all the external senses. In taste the sapid body, in smell the odorous particle, in hearing the sonorous vibration, and in sight the ray of light—must all *touch* the nervous part before sensation can be affected. The organs concerned in touch execute other functions besides, and in this respect touch differs from the other senses.

574. The skin of man, Fig. 41, p. 185, is peculiarly adapted for the exercise of the sense of touch, by possessing what is called a papillary apparatus, or a multitude of little paps in which the nerves of touch commence On the palmar surface of the hand and ends of the fingers, the papillæ are arranged in rows, and in these rows the papillæ are so numerous that microscopists estimate from one hundred and fifty to two hundred in a square line. They are almost equally numerous on the red surface of the lips. On other parts of the body where they are less numerous, the touch is less acute.

575. In the exercise of touch each layer of the skin executes a special office. The corium, or innermost layer, offers the necessary resistance when bodies are brought in contact with the surface ; while the rete mucosum, or second layer, by its pulpy consistence, keeps

the corium and the numerous sensitive paps on its surface in a delicate and supple condition. The insensible cuticle regulates the intensity of touch, and protects the papillæ from injury.

576. The degree of perfection in touch is greatly influenced by the state of the cuticle. Where it is thin, as upon the lips, the sense is very acute, but where the cuticle is thick and hard as upon the soles of the feet, the sense of touch is obtuse and uncertain. And if the cuticle be removed as by a blister, the sense is painfully acute and much impaired.

577. The only idea communicated to our minds by the sense of touch in its simplest exercise is resistance, and it is by the various degrees of resistance which the surface encounters that we estimate the condition of bodies which we feel. The relative sensibility of different parts of the skin may be judged of by touching the surface with the legs of a pair of compasses, the smallest distance at which the two points can be distinguished indicating the highest degree of sensibility. And in the trial of this experiment the tip of the tongue is found to possess the sense of touch in the highest degree, the two points of the compasses being distinguishable when only half a line apart. Next to the tip of the tongue, the palmar surface of the index finger is the most delicate; next, the palmar surface of the other fingers and the red surface of the lips.

578. It is chiefly in the variety of movements of which the hand is capable that the sense of touch in man is superior to that of the lower animals, and the sense of touch as thus employed affords us an important means of acquiring information and of correcting many vague falacies which we should otherwise incur from the sense of sight.

579. The clearness of our perception by touch is greatly increased when the tactile surface is made to carefully glide over the body which is being examined. In this way some blind persons so educate the sense that they are able to distinguish the *colors* of fabrics by the difference in the *grain* or *degrees of smoothness* imparted by different dyes.

580. TASTE, like touch, is excited by the contact of external objects upon certain parts of the surface, but it acquaints us with properties which escape touch, to wit, the *flavor* of substances.

581. The tongue is supplied with sensitive papillæ, like the skin, and it is difficult to distinguish those which transmit the sense of taste from those of touch. Besides, taste does not exclusively belong to the tongue, nor does it belong to every part of it, so that altogether the senses of touch and taste are intimately blended with each other.

582. For the correct appreciation of taste, it is necessary that the substance to be tasted be in a *liquid* state. Insoluble substances only convey imperfect sensations of touch, such as a feeling of cold. There are many substances to which we are in the habit of ascribing a hot, cold, acid, or caustic taste, which only excite the sense of touch without making any impression on the nerves of taste.

583. The lips, the inner surface of the cheeks, the gums, the palate, and the greater part of the anterior surface of the tongue are devoid of the sense of taste. While the quickest and most energetic perception belongs to the posterior half of the tongue, where the sense of touch is very imperfect.

584. The folds which surround the base of the tongue also possess the sense of taste, so that the mechanism

of swallowing affords the necessary movement for acquiring the taste of all soluble substances. And this is a beneficent provision in the object of taste, to direct us in the choice of food. Among the lower animals, the instinctive perceptions connected with taste are much more remarkable than our own. An omniverous monkey will seldom partake of poisonous fruit, although the flavor may be agreeable, and animals whose food is restricted to one kind will decidedly reject all other food. As a general rule, it may be stated that substances of which the flavor is agreeable to us are useful in our nutrition, and vice versa, but there are many exceptions.

585. Like the other senses, taste is capable of being educated. Wine-tasters and epicures cultivate this sense to a high degree of acuteness. And impressions realized by taste remain much longer than those by the senses of smell, sight, and hearing.

586. The impressions of some substances to the sense of taste are in part derived from their odorous emanations, of which we take cognizance through the sense of smell. It is well known that when the sensibility of the nasal membrane is blunted by catarrh, the power of distinguishing flavors is much diminished. Yet there are cases where the sense of smell has been wanting without any impairment of the sense of taste.

587. Smell is the sense by which we perceive the impressions made on the nerves by odorous particles suspended in the atmosphere. The nerves of smell only take cognizance of elastic or gaseous fluids, hence they are so distributed that the organ destined to receive odorous impressions must be so placed as to receive their contact ; and experience teaches us that to have the organ of smell discharge its functions, the membrane touched by the odorous particles must be

continually moistened and covered by a liquid adapted to fix the particles for a time and to absorb their odors. Smell must consequently reside in the walls of a cavity communicating externally, and the more rapidly and regularly the air conveying odors to us is renewed, so much the more favorable are the conditions for smelling.

588. The nose fulfills all the requirements here contemplated. It is lined by a soft velvety membrane, which, in health, is constantly lubricated with mucus, in this membrane the nerves of smell are distributed in innumerable filaments. When odorous particles diffused in the air are brought in contact with the lining membrane of the nose, as in the act of respiration, they are arrested by it long enough for the nervous filaments to take cognizance of their presence. From this it can easily be perceived, how the changes in the nature of the fluid secreted by the nose during a "cold in the head" may impair the sense of smell.

589. Sailors are familiar with the aroma from certain fertile places and countries in vegetable odorants which can be experienced at great distances. The perfumes of the Moluccas can be detected sixty miles. And it is recorded that the companions of Columbus were first made sensible of the approach of land by the odors several days before they discovered it.

590. Smell, which under ordinary circumstances is deemed the least important of the external senses to man, is of the utmost consequence to many animals. And the extent to which it may be cultivated by persons devoid of the other senses, shows it to be in many respects of equal utility with the rest. Daily experience shows that this sense is so inconceivably acute, in regard to otherwise incomprehensible quantities of odorous particles, that we can form some conception of

its great importance to the lower animals as a means of warning them of danger. And it is recorded of a man who, from peculiar idiosyncrasy, was liable to be made sick by the presence of a cat, detected one shut up in a closet near his room by the sense of smell, after repeated assurances by persons who had searched for it in vain, to satisfy what they considered a caprice.

591. Odors are sometimes so powerful as to produce stupefaction and fainting, others cause nausea and vomiting. And some odorants possess the quality of retentiveness to such a degree that they will remain for years attached to substances once imbued with them. A single grain of musk evaporated from a hot plate, will communicate to a dry room with papered walls an odor that will last for months.

592. SIGHT is the sense by which we distinguish colors and appreciate most of the qualities of external objects. It, too, is a modification of touch. And whether we regard light as an emanation from luminous bodies, or as the vibration of a subtle fluid, it is equally necessary for it to impinge upon the eye in order to convey the sense of sight.

593. The pleasure and the advantages derived by the mind through the sense of sight are of so signal a kind as to render the organ of vision a subject of universal interest. Every one who lays the slightest claim to a general education has made the eye more or less the object of study. It is not, however, my purpose to dwell upon the unspeakable simplicity and perfection of this organ, but to so display its parts as to give a clear comprehension of its function. To do this, we must have a correct appreciation of the mechanical properties of light.

594. A series of particles of light succeeding each

other in a straight line is called a *ray,* and the light which proceeds from a radiant point forms diverging *cones* of rays, which would proceed indefinitely did they not meet with obstacles.

595. The rays of light emanate with so great velocity from the sun, that they take only seven and a half minutes in traversing the immense space which separates the earth from that luminary. They travel at the rate of 192,500 miles in a second, and move through a space equal to the circumference of the earth in one-eighth of a second. They are propagated in straight lines so that they spread or diverge in all directions from the body whence they emanate, while their density diminishes in the direct proportion of the squares of their distance. So that, if the earth were at double its present distance from the sun, it would receive only one-fourth of the light; at three times its present distance, one-ninth; at four times its present distance, one-sixteenth, and so on.

596. In proceeding from a luminous body, the rays of light must traverse intermediate bodies in order to reach the eye. These bodies are called *media.* Air is the common medium, and when in this way the light has reached the exterior surface of the eye, the farther transmission is effected through different textures which form so many media.

597. When a ray of light impinges upon a body it is reflected, and bounds off, like an elastic ball, striking against the same surface in the same direction, would do, or it is absorbed and disappears; or, lastly, passes through the body, which, in that case, is transparent or diaphanous. In the first case, the body becomes visible, appearing white, or of some particular color, and we see it in the direction of which the rays reach the eye. In the second case, the body is invisible, no

light proceeding from it to the eye, or if the surrounding objects are illuminated it appears black. In the third case, if the body is perfectly transparent, it is invisible, and we see through it the object from which the light was last reflected. But light is subject to great changes in passing through transparent media.

598. A ray of light in passing through the same medium always takes a rectilineal course. But when it meets with a medium of a different kind, four different effects may result. Part of the light is dispersed on all sides, while another part is reflected in a regular or prescribed path. If the new medium be transparent, part will pass onwards in a refracted state. And, finally, part is absorbed or lost.

599. *Refraction* is the name given to the deviation of a ray of light from its original path in entering a medium of different density. The amount of refraction depends directly on the nature of the media traversed by the light. So that when a ray of light coming out of the air passes through a piece of glass, the result may be estimated by the index of refraction which is offered by glass in connection with the atmosphere.

Fig. 45.

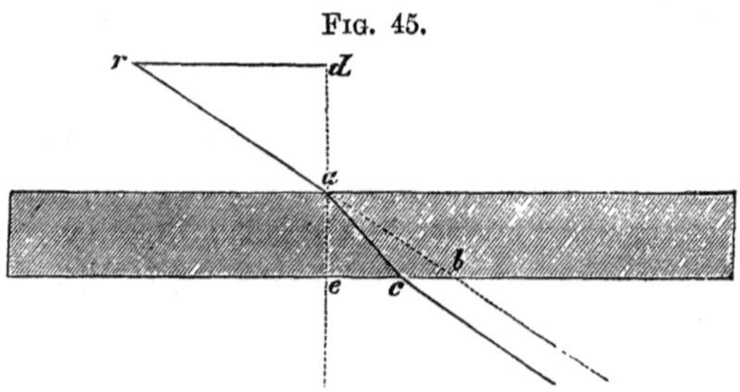

Refraction of Light.

600. Thus, when the ray of light *r*, passing through

the atmosphere, falls obliquely upon a plate of glass at the point *a*, instead of continuing to move in the same straight line *a b*, it is bent towards the perpendicular at *a*, and proceeds in the direction *a c*. The ray is bent to the side in which there is the greatest mass of glass. On emerging from the glass into the air, a rarer medium, at the point *c*, the ray has its direction again changed, and in this case *from* the perpendicular, but still towards the plate of glass.

601. The amount of refraction is generally proportionate to the density of the body, but combustible bodies possess a higher refracting power than corresponds to their density. Hence, the diamond, naphtha, melted phosphorus, and some other substances, exhibit this effect upon light in a greater degree than other transparent bodies. The same substance, however, always has the same refractive power, whatever may be its shape. The degree of refraction of light in traversing a single transparent surface is measured, by comparing the obliquity of its approach to the surface with the obliquity of its departure on emerging, and for this purpose a line is supposed to be drawn through the surface at the point where the ray impinges as at *a e*, and the relative positions of the ray to this line on both sides of the surface are easily ascertained. Thus the line *r d*, drawn from any point of the ray before passing to such a perpendicular, is a measure of the original obliquity of the ray, and is called the *sine of the angle of incidence*, and the other line *a e*, drawn from a corresponding point of the ray after passing to the perpendicular, is the measure of the obliquity *after* refraction, and is called the *sine of the angle of refraction.* By comparing these two lines in any case, the amount of the difference constitutes the degree of refraction.

602. The *angle of incidence*, *r a d*, is the angle formed

by the incident ray *a*, with the perpendicular raised from the point of immersion; the *angle of refraction*, *a e c*, is that formed by the refracted portion of the ray with the same perpendicular.

Fig. 46.

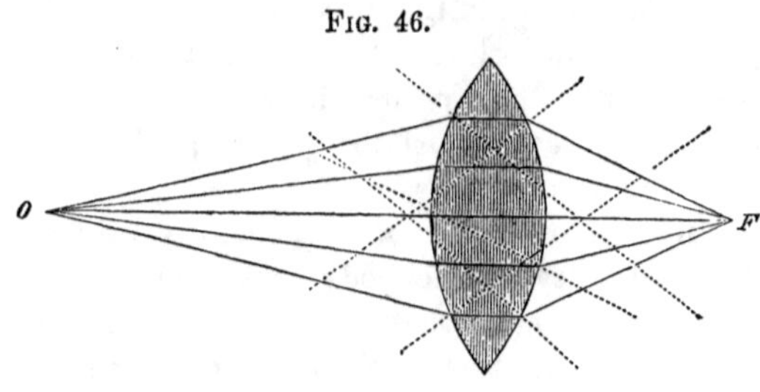

603. If the surface of the medium be convex, Fig. 46, the rays are so situated, after refraction, as to converge behind the refracting body into a point called a *focus*, *F*, which is nearer to the medium the less the divergence of the rays, or, in other words, the more distant the luminous objects, *O*.

604. But if the surface of the medium be concave, Fig. 47, the luminous rays proceeding from the object, *O*, are rendered so divergent as to form a focus, *F*, anterior to the medium.

605. A knowledge of the various effects which different transparent media produce on rays of light has led to the construction of numerous invaluable optical instruments adapted to modify the rays of light, so as to change the situation in which bodies are seen; to augment their dimensions, and to render them

Fig. 47.

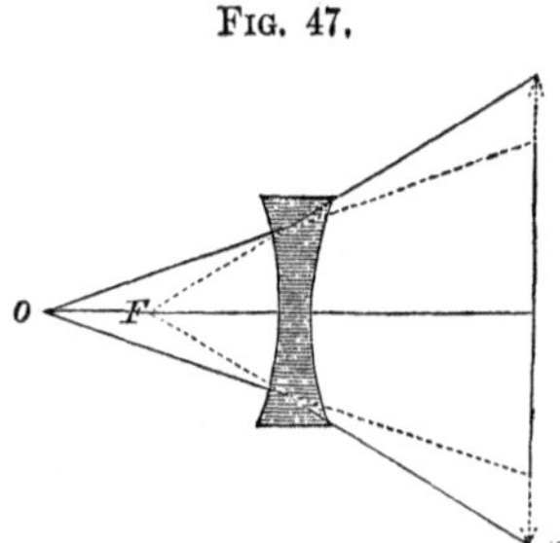

more luminous and visible, when remote and minute. The simplest instrument used for this purpose is a double convex glass shaped like the bean of a lentil plant, and hence, the name of *lens*, which is applied to variously shaped bodies used for the same purpose.

FIG. 48.

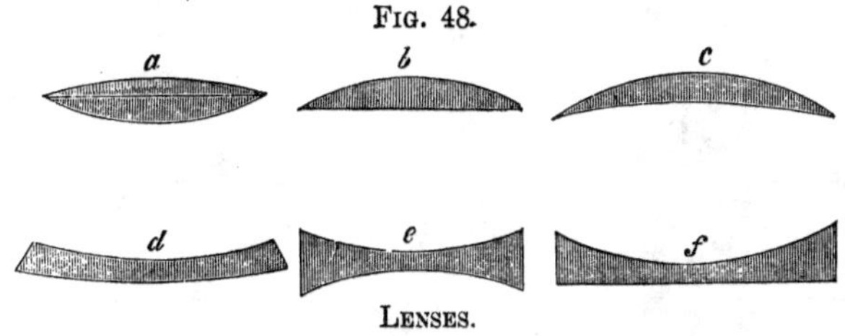

LENSES.

a, double convex ; *b*, plano-convex ; *c*, meniscus, or moon-shaped ; *d*, concavo-convex ; *e*, double concave ; *f*, plano-concave.

606. The lenses *a*, *b*, and *c* are called collective, because they collect the originally divergent rays in such a manner that the latter unite behind them. On the other hand, *d*, *e*, and *f* form *dispersive* lenses, since they tend to increase the divergence. The foci of the collective lenses are said to be *real*, because they lie on the opposite side of the luminous object, while the foci of dispersive lenses lie on the same side with the object, and are called *virtual* foci.

607. Another element in the change which takes place in light by the power of refraction is the *different colors* which it exhibits, and the difference in the refractability of these colors.

608. It is only by the decomposition of light into its constituent colors that we are enabled to explain the cause of the different shades of color. When a white light impinges upon a body, the body either absorbs all the rays that compose it, reflects all, or absorbs

some, and reflects others. If it reflects the whole of the light to the eye, it is of a white color; if it absorbs all or reflects none, it is black; if it reflects only the red ray, and absorbs all the rest, it is red. And so of the other colors.

FIG. 49.

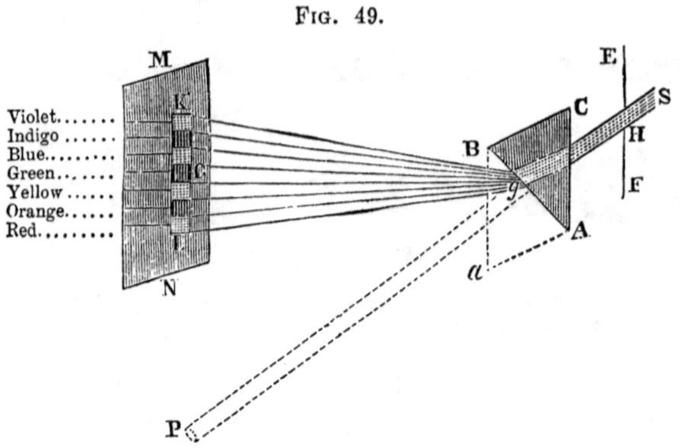

609. If a ray of sunlight be admitted through a small hole *E F*, into a dark chamber, it will proceed in a direct line to *P*, and form a white spot upon the wall. But if a glass prism, *B A C*, be placed so that the light may fall upon its surface, *C A*, and emerge at the same angle from its second surface, *B A*, in the direction of *G*, the ray will expand; and it will be found to occupy a considerable space, *M N*, and instead of the white spot there will be an oblong image of the sun, *K L*, consisting of seven colors, namely, red, orange, yellow, green, blue, indigo, and violet. The whole of this group of colors has received the name of *solar spectrum.* It was first depicted and explained by Newton. These colors are incapable of further decomposition, while they are found to possess different degrees of refrangi-

bility. The red rays experience the least, and the violet the greatest refraction.

610. Upon the differences in the power of appreciating colors by different persons, depend the excellences of harmonious coloring by painters. If, after gazing intently on a luminous object, we suddenly look upon a white ground, we shall perceive a dark spot upon it; the portion of the retina which had been impressed by the bright image not being immediately capable of an impression from a less luminous object. The apparent spot on the white ground, as it fades away, seems to go through a succession of shades; these shades indicate the changes that take place on the retina in its return to the natural condition, and their order of succession indicates the *complementary* color to that last appreciated. For example, if we look intently for a length of time upon a bright red object on a white ground, and then suddenly evert the eye so as to look upon the white ground only, we see a green spectrum, and this is the complementary color to red.

611. All complementary colors have an agreeable effect when judiciously combined, and all bright colors if not complementary have a disagreeable effect. This is particularly the case in regard to the primary colors, strong combinations of any two of which, without any color that is complementary to either of them, are displeasing.

612. Painters who are ignorant of the physiology of light, usually introduce a large quantity of dull gray into their pictures, so as to diminish the glaring effects which they would otherwise produce. But the advantage obtained by this is at the sacrifice of that vividness and force which would be secured by a harmonious combination of complementary colors.

613. Some persons, however, are incapable of correct discrimination of colors, even though their powers of vision in all other respects are perfect, and there are rare instances where the defect is in relation to some one or more particular colors. This defect is commonly called Daltonism, from the name of a distinguished philosopher who was the subject of it, and who was the first to describe it. Dalton was unable to distinguish *blue.*

614. *Spherical aberration,* or of that deviation of light which is due to spherical shape, is also intimately connected with the sense of sight

FIG. 50.

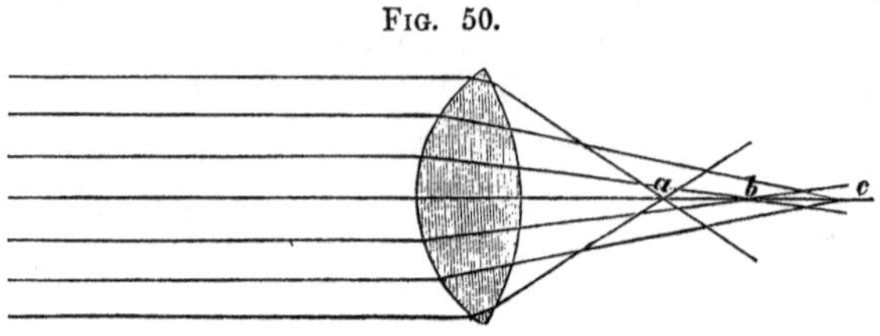

615. Let us suppose that in this figure (Fig. 50) of a double convex lens, the line *a b c* occupies its prolonged axis ; the several parallel rays will have different chief foci, and those rays which penetrate the lens at a distance from its axis unite behind it sooner than those which reach the lens nearer its centre. Thus the chief focus of the former is at *a ;* while that of the latter is placed at *b* or *c.* This proposition applies to all rays that proceed from a luminous point, which is too near to allow of their being regarded as parallel. The distance *a b* and *b c* corresponds to the amount of spherical aberration. This will necessarily render the image of the focus more obscure, since it will prevent the

rays from intersecting in the same transverse plane to form it.

616. When the rays are not far from the prolonged axis the spherical aberration is but small, and may be disregarded. And in a lens that has slight curves with large diameters, the difference of the angles of the external and middle divergent rays is diminished. Hence, lenses of this form are frequently used by opticians, in order as much as possible to avoid spherical aberration.

Fig. 51.

617. But when it becomes necessary to use a lens which offers on the whole a considerable spherical aberration, it may be diminished by the use of a septum or diaphragm (Fig. 51). A partition with a perforated centre will admit the central rays which have no spherical aberration worthy of notice, while the outermost rays are completely excluded.

Fig. 52.

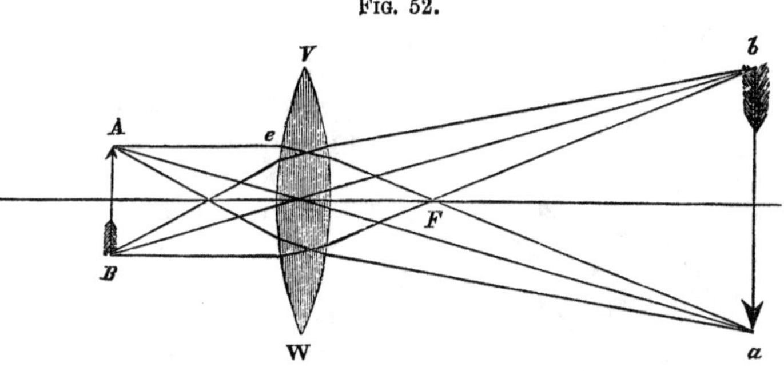

618. Let $V\ W$, Fig. 52, represent a double convex lens, having its chief foci at F. If we place an object, $A\ B$, at from once to twice the focal distance, we shall

get a real, inverted, and magnified image at *b a*. When the object lies at more than twice the focal distance, or so that *a b* may be regarded as the image from which the rays of light proceed, we also get a real, inverted, but diminished image, *A B*.

Fig. 53.

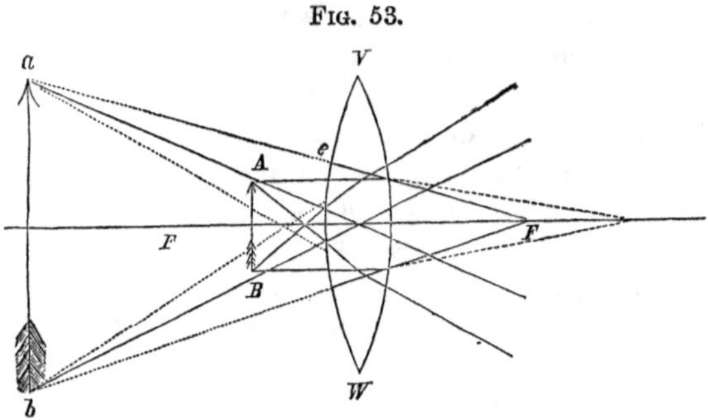

619. Again: if the luminous body, *A B*, stands between the principal focus, as at Fig. 53, and the lens, *V W*, we obtain a virtual, upright, magnified, and more distant image, *a b*.

620. A *Camera Obscura* consists of a box, *a b c d*, Fig. 54, which is blackened internally, and has a posterior surface fit for the reception of images, such as is formed by the plate of ground glass, ☞. Its tube, *e*, has anteriorly a double convex lens, *f g*, which produces a suitable refraction of the rays. When an object is placed at more than twice the focal distance of *f g*, we see upon ☞ an in-

Fig. 54.

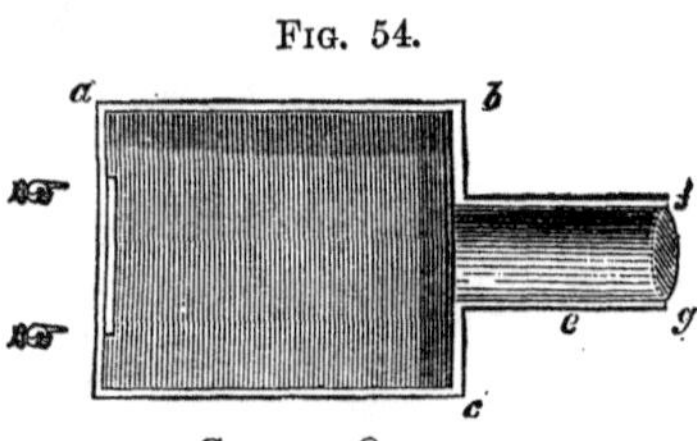

Camera Obscura.

verted and diminished image, the size of which constantly decreases with an increasing distance of the luminous point. But since the focal length varies with the distance of the object, the tube, *e*, must be capable of being lengthened or shortened, so as to allow of the situation of the lens being altered in a corresponding degree. We are thus enabled to adjust the focus of the camera to different distances. The greater the absolute distance of the luminous body, the smaller the differences of focal length become. This explains why, for all ordinary purposes, comparatively slight alterations of *e* will suffice.

621. The *telescope* which may be regarded as a camera obscura on a large scale has, with much propriety, been compared to the eye, as many of the parts of that instrument have been added to execute particular offices which are performed by the eye—the most perfect of all optical instruments. Every telescope consists, in part, of a tube, which always comprises pieces capable of adjustment to each other. Within this tube several glasses or lenses are so arranged as to refract the rays of light and bring them to a determinate foci. And within the telescope is a partition having a round hole in its centre, usually placed near a convex glass or lens, with the object of limiting the amount of light to be admitted, and obviating spherical aberration; and that this purpose may be more perfectly accomplished, the interior of the tube and the diaphragm or partition are colored black, so as to be capable of absorbing the oblique rays which are not conducive to vision, and which would otherwise cause confusion. The whole arrangement is nearly a counterpart of that which exists in the eye.

622. The EYES are so placed as to command a great

extent of vision at one time, while they are protected in two bony cavities, known by the name of orbits. Above and around these are arranged the eyebrows and eyewinkers, with their more or less dark color, for the purpose of softening the rays of light ere they impinge upon the more sensitive globe, and to shade the eyes from the continued impression of light. And the no less useful tears incessantly flow over the surface, preventing the effects of friction which would result from a dry membrane, and also for the purpose of defending the eye against injury from particles of hard bodies floating in the atmosphere.

FIG. 55.

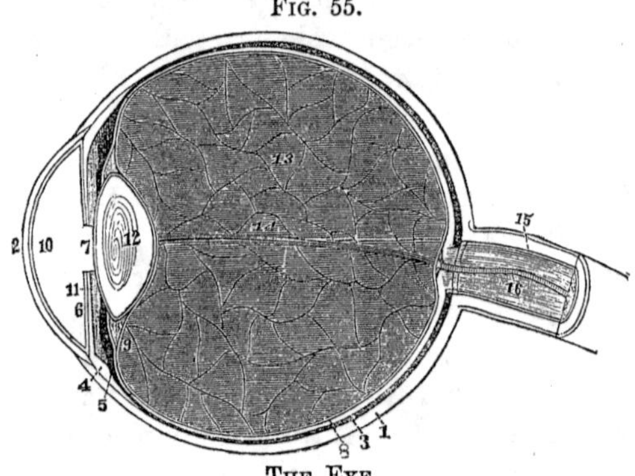

THE EYE.

A longitudinal section of the globe of the eye. 1. The sclerotic coat thicker behind than in front. 2. The cornea received within the anterior margin of the sclerotic, and connected with it by means of a beveled edge. 3. Choroid, connected anteriorly with the ciliary ligament 4, and the ciliary processes 5. 6. The iris. 7. The pupil. 8. The retina, or third layer, terminating anteriorly by an abrupt border at the commencement of the ciliary processes. 9. The canal of Petit, which encircles the crystalline lens, 12. 10. The anterior chamber of the eye. 11. The posterior chamber. These chambers are separated by the iris, and contain the aqueous humor. 12. The crystalline lens, more convex behind than before, and inclosed in its proper capsule. 13. The vitreous humor inclosed in the hyaloid membrane. 14. A tubular sheath of the hyaloid membrane, which serves for the passage of the artery of the capsule of the lens. 15. The optic nerve. 16. The central artery of the retina.

623. The globe of the eye is attached to the optic nerve as to a handle, and is imbedded with it in a cushion of fat contained in the orbit. Its shape is that of a hollow sphere, a little protuberant in front; and it is entirely filled with special refractile substances. Its exterior envelope consists of three membranes—the *sclerotic, choroid,* and *retina;* the last being the one that receives the impression of light. Within are placed in succession—from before backwards—the *cornea, aqueous humor, crystalline lens,* and *vitreous humor,* and, besides these, across the interior of the eye, near the anterior surface of the crystalline lens, is a diaphragm—the *iris,* having an aperture in its centre, the *pupil.*

624. The *sclerotic* or outer coat, so named on account of its hardness, is that which gives shape to the organ, and constitutes the white of the eye. Immediately in front of the eye, completing the sphere of the sclerotic, is the transparent *cornea,* so called on account of its horny consistence. Its edges are fitted to the sclerotic coat, like a watch-glass applied upon a sphere, and projecting beyond the surface. The second coat or *choroid* (having relation to color) is formed by a very thin network of blood-vessels stained with a black fluid, named *pigmentum nigrum;* it is this substance which gives the bottom of the eye that deep color seen through the pupil. The absence of the pigmentum nigrum constitutes an *albino.*

625. The choroid coat, like the sclerotic, is a segment of a sphere, terminating near the margin of the cornea. At its termination, the margin is plaited into a number of little triangular bodies—the *ciliary processes*—which surround the lens like the corolla of a flower.

626. The *iris* is the thin colored verticle septum placed before the aperture in front of the choroid, which it assists to close. It is retained in its place by means of the *ciliary ligament* —a name given to a little circular ridge or seam—at the junction of the external margin of the iris with the folded edge of the choroid. By the position of the iris across the space included between the cornea in front and the lens behind, it divides this space into two apartments called *chambers*. In the middle of the iris is an aperture—the *pupil*—by means of which the two chambers communicate, and by it the rays of light are admitted to the retina, the last or internal coat.

FIG. 56.

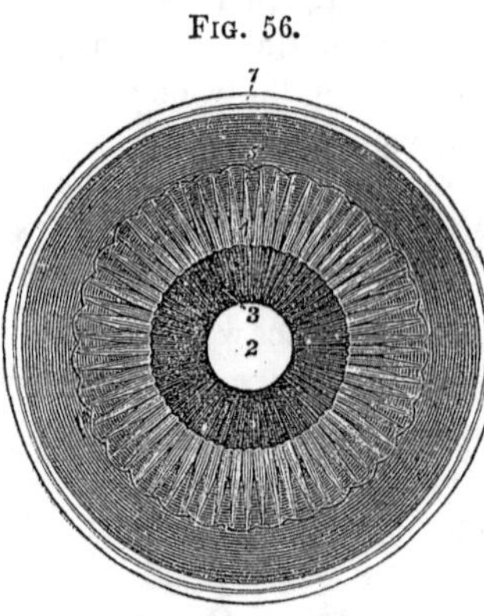

THE IRIS.

Anterior segment of a transverse section of the globe of the eye, seen from within. 1, the divided edge of three tunics ; 2, the pupil ; 3, the iris ; 4, the ciliary processes ; 5, the scalloped anterior border of the retina.

627. The *retina* is the seat of vision, it lines the choroid, and is a soft, dim, pulpy, and grayish membrane formed by the expansion of the optic nerve. It is also a segment of a circle terminating in a scalloped border at the ciliary ligament. Corresponding to the entry of the optic nerve and in the direction of a line drawn perpendicularly through the centre of the cornea, is a yellow spot named *limbus luteus*, about a line in extent, which is supposed to be the most sensitive portion of the retina.

628. The *crystalline lens* is a transparent double convex body, situated immediately behind the pupil and in front of the *vitreous humor*, so that all the rays of light that pass through the pupil, must traverse it, by which, with the other structures they are refracted to depict the image on the retina. The posterior surface of the lens is generally more convex than the anterior, but this differs in individuals, and at different ages the shape of the lens presents great varieties. Its convexity decreases with increasing age, and in old people it is flattened, occasioning *presbyopia* or long-sightedness. Whereas when the lens is excessively convex, the result is *myopia* or short-sightedness.

FIG. 57.

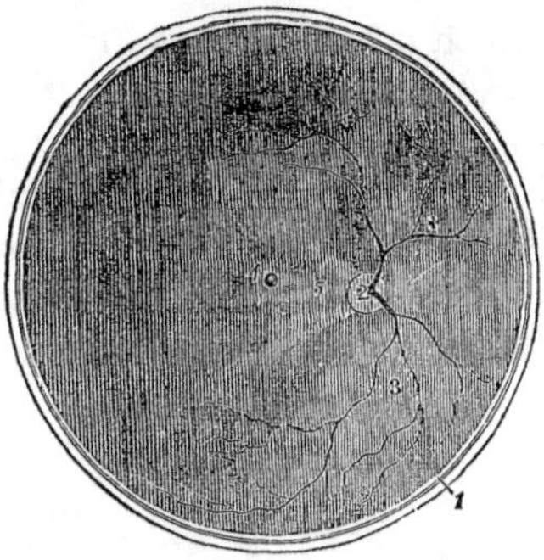

THE LIMBUS LUTEUS, OR YELLOW SPOT OF THE RETINA.

1, divided edge of the three tunics of the eye transversely, showing the posterior segment from within; 2, the entrance of the optic nerve with the artery of the retina piercing its centre; 3, 3, the ramifications of the artery; 4, the LIMBUS LUTEUS, with a little hole in the centre, showing a deficiency of the retina at this point; 5, folds of the retina, which, when the eye has been opened generally obscure the hole in the centre of the limbus luteus.

629. The crystalline lens is retained in its place by means of a thin, firm, and elastic transparent membrane called a *capsule*, which closely surrounds it, and at the margin of the lens, where the two sides of the membrane unite, there is a little circular cavity called after its discoverer, the *canal of Petit.* The capsule is connected to the lens and at its edges with the ciliary

process by the arteries which perforate and ramify all the tunics.

630. Of the *fluids* of the eye, the *aqueous humor* is that which is contained in front of the lens and fills the chambers. The quantity of this fluid is about four grains, and its specific gravity is very near that of water. The surfaces of the chambers are lined by an exceedingly delicate transparent membrane, which contains the aqueous fluid. The *vitreous humor*, or *body*, is so named on account of its resemblance to molten glass. It supports the delicate structure of the retina which is applied to it, and entirely fills the cavity behind the lens, about three-fourths of the globe. In front the vitreous body is hollowed out to receive the convex surface of the lens to whose capsule the *hyaloid membrane* (from *ualos*, glass, and *eidos*, resemblance), which contains the vitreous humor, is adherent. The vitreous humor consists of water holding in solution albumen and soda ; it is perfectly transparent, and has a specific gravity rather greater than water.

631. The transparent parts of the eye are thus seen to be of different densities, and are, consequently, possessed of different refractive powers.

632. It is obvious that the rays of light which fall upon the transparent cornea can alone be conducive to vision. Nor, indeed, does the whole of the cornea admit light, for it is commonly covered in part, by the free edge of the eyelids. Those rays which impinge upon the sclerotica, and a part of those which fall upon the cornea, are reflected and produce the brilliancy of the eye with the image observed at its bottom. Besides, all the rays that enter the cornea do not impinge upon the retina, some fall upon the iris and are reflected back in such a manner as to indicate the color of the eye. It is

in short, only those rays of light which *pass through the pupil* that fall upon the retina and give rise to the sense of sight.

FIG. 58.

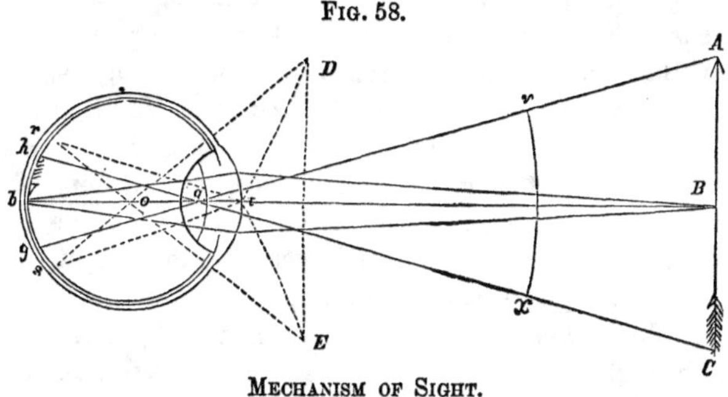

MECHANISM OF SIGHT.

633. Let us suppose a luminous cone to proceed from a radiant point *B*, Fig. 58, directly in the prolongation of the antero-posterior diameter of the eye, the axis of this cone will also be the axis of the organ, and a ray of light impinging upon the humors in the direction of the axis will pass through them without deflection, and fall upon the retina at *b*. But this is not the case with the other rays composing the cone. As they do not fall perpendicularly they are variously refracted in their passage through the cornea, aqueous humor, crystalline lens, and vitreous humor, yet, in such a manner that they join their axis in a focus at the point where it strikes the retina.

634. A ray of light in proceeding through the aqueous humor is subject to but little variation, as the density of it differs but little from the cornea, this latter, however, is rather more refractive, and the little tendency that exists will be to render the ray less convergent. This convergence makes provision for the

entrance of a greater number of rays through the pupil and necessarily intensifies the light that impinges upon the lens. Following the ray through the two chambers of the eye, we find it next impinging on the surface of the lens, which possesses a much higher refractive power than the cornea or aqueous humor, the ray is, therefore, more powerfully converged and brought nearer the axis of the cone. Passing through the crystalline lens, the ray emerges into the vitreous humor—a medium of less refractive power than the lens—which again deflects it and increases the magnitude of the image to be depicted upon the retina.

635. When rays fall obliquely, *A*, *g*, and *C*, *h*, and pass through the centre of the lens, they are refracted at each of its surfaces, but as the two refractions are equal and in opposite directions, the rays may be regarded as pursuing their course in a straight line. As, however, the oblique ray has to pass through the cornea and aqueous humor before it impinges on the lens, it undergoes considerable deflection by these substances, although the main deflection occurs at the entrance of the rays into the cornea, as seen at *D*, *t*, *s*, and *E*, *t*, *r*, of Figure 58. The point *a*, at which the various rays cross, is called the *optic centre* of the lens. Each of the straight rays proceeding from a radiant point may be assumed as the axis of all the rays proceeding obliquely from the same point, and the common focus must fall on some part of this axis. In this way the object is represented in miniature and *inverted*, on the retina—the rays from the upper part of the object falling upon the lower border of the retina, and those from the lower part falling upon the upper border of the retina. Hence it becomes an important question, why objects are not perceived in an inverted position.

636. Many philosophers have thought all attempts to explain the cause of erect vision as merely speculative, and therefore superfluous. But such persons would limit their reasoning to the organ of sight, regarding it simply as an optical instrument capable of being imitated and worked by the hand. Others attempt an explanation by supposed analogies; such as, that the rotation of the earth is not perceived when we look at the different bodies on its surface, since we and all other masses are alike continually carried onward, so that all relative change of place is absent, and only becomes manifest when we attempt to compare our position with that of the sun or a fixed star, which does not share the movement. And, similarly, since all bodies are seen inverted, there are no relative differences which can betray the error. Opposed to such a view is the sense of touch, by which we recognize surfaces as upright, and that the necessary movements of the eye correspond to the true and not to the inverted position.

637. Others suppose that we gradually learn that it is necessary to direct the gaze upwards in order to see the top of an object, and downwards to see the base. This explanation presupposes the first realization of the sense of vision as seeing objects inverted, or it implies an indifference to the position of the image on the retina. And if it were true, until education was completed in this respect, there would be a conflict between the sight and touch. Experience proves this supposition erroneous; for persons who have been born blind, and afterwards obtain their sight, see objects at once in the erect position.

638. Our appreciation of an object by the sense of sight is not brought about by the simple transfer of an

impression upon the retina, like the reflected image of a mirror ; but it is an *impression of the mind,* excited by the image and depending upon it as the effect of a cause. And, forasmuch as there is in the eye a *centre of direction* through which all rays of light necessarily pass ere the object is depicted upon the retina, it is surely far more reasonable to suppose that the object suggested to the mind by the retinal image should be in an *erect,* rather than in an inverted position. Besides, as all the rays both transversely and vertically cross each other at the centre of direction, our sense of the relative position of objects, or of the different parts of the same object, is derived from a mental appreciation at every point of impingement upon the retina, and in a line which joins it to the centre of direction for the whole object reflected.

639. Our judgment of the superficial extent of any object depends upon the size of its image on the retina. We notice how much is seen distinctly, and how much indistinctly, by the proportion of the rays which fall upon the *limbus luteus* or yellow spot, and those which fall on the retina outside of it ; and we refer the several neighboring points of the image to a certain distance, and conclude from hence the superficial extent of the object. But as we have no means of accurately measuring the distance of luminous objects by sight, we frequently make great mistakes in our estimates by vision.

640. *Single vision* with the *two* eyes has been the subject of scarcely less discussion than the cause of seeing objects in the erect position. Speculation on this, like that on the attitude of bodies, turns upon the difference between philosophical and physiological deductions ; the former usually disregarding the *single*

mental appreciation of impressions simultaneously suggested by two pictures of the same object.

641. We have seen that those points of a luminous body are most plainly perceived, which are mirrored upon the retina at the centre of the visual axis, and that those rays which fall outside this place gradually increase in obscurity, and finally fail to make an impression. Hence, all satisfactory perception is limited to an angle of certain width at the optical centre. Supposing this angle to be *v x* (Fig. 58), it is obvious that the extent which can thus be looked over will increase with the distance. Hence, while a healthy eye can only look over the arc of a circle comprising a few feet in its immediate neighborhood, it can include an arc of many miles if the circle be sufficiently increased.

642. The movements of the eye depend upon a set of muscles, which rotate the perceptive surface around the centre of the organ. And by having *two* eyes with a synchronus muscular arrangement, the individual is spared many complicated adjustments which must have been necessary to adapt one eye to the function of distant vision. The horizontal arc of movement producible by the muscles of the eye amounts to about 110°, and the perpendicular to about 100°. By trying the experiment of turning the eyes as far as possible to either side, it will be found that the limits of possible perception by the two eyes may be completely separated from each other, neither one being able at the ordinary reading distance of looking more than a few inches towards the other, while the everted eye turned in the same direction is capable of looking over the shoulder. When both eyes are directed forwards, the visual curves intersect each other, and thus form a single centre. And when luminous objects throw their images upon

the sensitive portions of the retina, the two together form a common visual circle.

643. Singleness of vision in an object that is looked at with both eyes, depends upon the intersection of the rays of light by the convergence of the visual circle. This may be demonstrated by arranging two objects in the line of convergence, when they will appear as one. This is accomplished by looking through two tubes placed before the right and left eyes respectively, at two similar objects of any kind placed near the farther end of the tubes. And if the objects be slightly approximated so that the axis of the tubes meet in a more distant point, the mind is impressed with the image of only a single object, and this appears to be removed back to the point of convergence.

644. We thus discover that the object of two eyes is not that of enabling us to perceive two different views simultaneously, but to render a single observation clearer, rather than to confuse an instantaneous act by numerous complications. Besides, when both eyes are employed they concur in exciting the perception of solidity or projecting surfaces, which arises from the mental combination of the *two* pictures formed upon the retina, which in the use of one eye only, is never apparent for new objects. The appreciation of solidity by a person who has never had the use of but one eye, depends upon the combination of the sense of touch with that of vision. A case in point is recorded by Cheselden :—A youth, born blind, at the age of twelve years, gained the sight of one eye by a surgical operation ; and for some time after he had tolerably distinct vision, everything appeared to him *flat*, as in a picture. And it was not until after he had acquired the power of judging of the real forms and distances of the objects

around him by the sense of touch that he began to have a correct appreciation by sight. It is said of him that he was well acquainted with a dog and a cat by *feeling*, but he could not remember their respective characters when he *saw* them. And one day when thus puzzled, he took up the cat in his arms, and feeling her attentively so as to associate the sense of touch with that of vision, put her down, saying: "So puss, I shall know you another time." After much education by the sense of touch, he fell into the converse error of supposing that a picture, which was shown him, was the real object represented in relief on a small scale.

645. It is, indeed, capable of proof that it is only by the mental association of two pictures upon the retinæ that we can derive any idea of solidity by the sense of sight. To demonstrate this, Professor Wheatstone invented and first described the stereoscope, in 1838.

646. The *stereoscope* essentially consists of two plane mirrors, inclined with their backs to one another at an angle of 90°. If two perspective drawings of any solid object be so placed before these mirrors, one before each, that their two images shall be made to fall upon the corresponding parts of the two retinæ, in the same manner as the two images formed by the solid object itself would have done, the mind will receive a single projecting or solid surface, the exact counterpart of that from which the drawing or photograph was taken.

647. To clearly appreciate the mechanism of sight in the stereoscope, we will disregard the interposed mirrors, and assume that the drawings lie immediately in front of the eye, but with the same relations as their reflected images, each of which sends its rays to the requisite part of the retina of the corresponding eye. Supposing that the prolonged axes of vision *c o* and *f l*

FIG. 59.

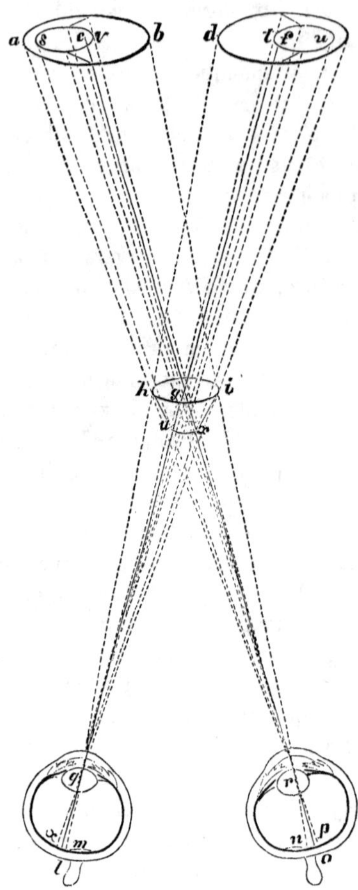

impinge upon the two centres of the greater circles, *c* and *f*, these are perceived as a single point, *g*, at the place where the conducting lines, *c o* and *f l* intersect each other. But *a* has the line of direction *a p*, and *d* that of *d m;* and both meet identical places of the retina, *m* and *p*. Hence, they are also seen single at the intermediate point, *h ;* and for the same reason *b* and *e* appear at *i*. Since this is repeated at every point, we get the single circle *h i*, having *g* as its centre. Applying the same construction to the smaller circles, we shall also find a median circle, which will be in front or behind that of the base, according as the eye perceives the single drawing designed for that effect or the opposite one. The intermediate lines give rise, point by point, to corresponding lateral lines. In this way, the view in space of either truncated cone, *h i x u*, may be constructed geometrically.

648. If the ranges of vision, and the influence of close observation, remained limited to a single space, or even to the intersection of the conducting lines, the other parts of the solid cone would be less distinctly perceived. Careful examination teaches that this really is the case. And something similar to this occurs in the perception of all solid objects.

649. *Distance* is appreciated in the same manner as solidity, namely, by the mental combination of the perceptions derived from pictures upon the retina of *two* eyes. How much our estimation of the relative distances of near objects depends upon the use of both eyes, may be made evident by closing one, and trying to execute any simple purpose requiring exactness, such as threading a needle. Of *remote* objects, our judgment of distance is chiefly founded upon their apparent size, when this is known to us ; but if such is not the case, our knowledge of the intervening space is chiefly deduced from the color and outline, or what is known to artists as "ærial perspective."

650. The *limits* of distinct vision greatly differ in different persons and at different ages, yet there is a maximum and a minimum distance at which a luminous body can be plainly seen by most persons, whatever may be the focal distance of their eyes. It is possible, by a bright light and the closest attention, to recognize magnitudes between 1-405th and 1-540th of an inch. Particles smaller than these cannot be discerned with the unaided eye when single, but may be seen when placed in rows. Particles which powerfully reflect light, such as gold-dust, may be seen when not more than 1-1100th of an inch. And *lines* or opaque threads are visible when only about half the size of the silk-worm's fibre, or 1-4900th of an inch in diameter.

651. The chief difference in individuals as regards the limits of vision depends upon the degree of attention which they may be able to use. Many persons being able to distinctly see objects when pointed out, who cannot otherwise distinguish them, and the same is the case with distant objects. An object once made out, however, may be diminished by distance and easily retained in view, which, to begin with, involves the closest possible attention.

652. Hearing.—*Sound* is the sensation we experience whenever the vibrations of an elastic body strike our ears. Unlike light, sound has no distinct existence ; the sonorous rays are merely vibrations of air produced by the percussion of a sonorous body. All bodies are capable of producing it, provided their molecules are susceptible of a certain degree of reaction and resistance. When any such body is struck, the integrant particles experience a sudden concussion, which causes them to oscillate with more or less rapidity ; and this tremulous motion is communicated to the

bodies applied to its surface. If we touch a bell that has been struck by its clapper, we feel a certain degree of trembling. The air which surrounds it, being exceedingly elastic, receives and transmits the vibrations, which, coming in contact with our ears, convey to us the sense of sound.

653. Sound has been analyzed as well as light, and the use of the ear with regard to sound corresponds to that of the prism with regard to light. And the modifications of which sound is capable are as numerous and as various as the shades between the primitive colors.

654. The *force* of sound depends entirely on the extent of the vibrations experienced by the molecules of the sonorous body. But like light, it may be increased by collecting and concentrating its rays, or reflected by a resisting body. A notorious instance of collecting sound is recorded of the tyrant Dionysius, who had the roofs of the dungeon at Syracuse so formed, as to collect the words and even whispers of the prisoners ; and to direct the sounds, he had a hidden conduit extending from the place where he sat listening, to the centre of the concave roof. The *ear-trumpet* is constructed on the same principle as Dionysius' dungeon. A wide concave end to concentrate the vibrations connected with a tube to conduct them to the ear.

655. *Echo* is reflected sound. When the ærial oscillations meet with a resisting body of regular surface, they are reflected at an angle equal to the angle of incidence ; consequently, sounds heard in the course of reflected waves always deceive us as regards the direction of the sonorous body. If, however, we are so placed as to receive the direct oscillations in one direction and the reflected from another, we then hear both

the sound and the echo. Echoes may be repeated according to the number of reflecting bodies.

656. The *velocity* of sound may be easily estimated. The report of a cannon fired at a distance within the range of vision, is only heard *after* the eye has perceived the flash. And, considering the light to have reached the eye instantaneously, if we note the time that elapsed between the appearance of the flash and the report, and measure the distance of the gun, we can ascertain the rapidity of sound. This is found to be about eleven hundred and forty-two feet in a second. Knowing the velocity of sound, we can tell the distance of a thunder-cloud, by noting the time between the lightning-flash and the thunder-clap. If it be fifteen seconds, the cloud is at the distance of fifteen times eleven hundred and forty-two feet, or of three miles and a quarter. Fluids and solids are also conductors of sound, and convey it much more rapidly and perfectly than air. In water, the velocity has been estimated to be about four thousand nine hundred feet in a second, or more than four times as great as in air. And as the sense of hearing is exceedingly acute in fishes, the facility with which sound is conducted through the water, is to them a great source of protection. Solids conduct sound with still greater velocity than fluids, and with much more accuracy than either air or fluids. If the end of a long wooden rod is held in contact with the ear, the slightest scratch of the farther extremity may be distinctly heard. A knowledge of this fact and of the facility with which sound is conveyed through solids, lead to the discovery of the *stethoscope*, an instrument which physicians use to ascertain the condition of internal organs, by the sounds they produce.

FIG. 60.

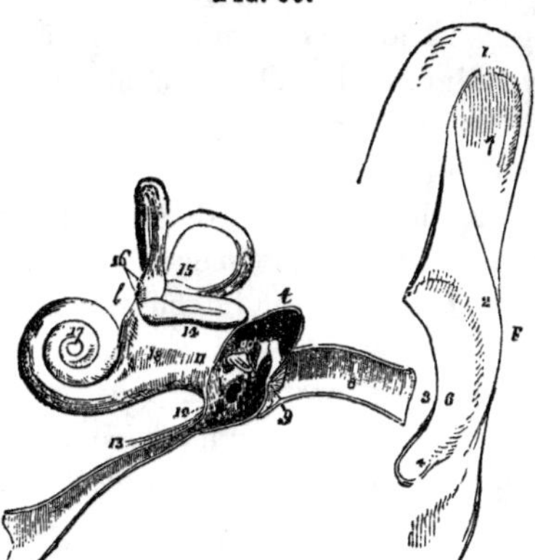

THE EAR.

Section of the human ear. *p*, the pinna ; *t*, the cavity of the tympanum ; *l*, the labyrinth ; 1, the helix (a fold) ; 2, the antihelix (opposite the fold) ; 3, the tragus (signifies goat, because this part is sometimes covered with hair like that of a goat) and the little lump opposite this, 4, is the antitragus ; 5, the lobulus ; 6, the concha (the name of an ancient measure which held half an ounce) ; 7, the upper part of the fossa innominata or nameless cavity ; 8, the meatus or duct ; 9, the membrane of the tympanum or drum, divided by the section ; 10, the *three* bones, malleus (hammer), incus (anvil), and stapes (stirrup) ; 11, the promontory, a bony elevation ; 12, the fenestra rotunda, or round window, this opens into a cellular net-work of hard bone behind the ear, called the mastoid cells ; 13, the Eustachian tube, so called from Eustachius, its discoverer ; the line on this represents a little muscle stretching across from the Eustachian tube to be inserted into the handle of the malleus ; 14, the vestibule ; 15, three semicircular canals, horizontal, perpendicular, and oblique ; 16, the ampullæ (from ampulla, a jug) ; 17, the cochlea (a snail shell) ; 18, the convexities of the two tubes which communicate with the tympanum and vestibule.

657. The EAR presents such a combination of all those physical qualities pertaining to æriform, liquid, and solid bodies, as are best calculated to realize the varying influences of sonorous bodies. The nature and

the importance of the function which is the source of so much delight in all the occurrences of life, will render the ear an interesting object of study to all who would read in this wonderful piece of mechanism the attributes of its Author.

658. The ear consists of three distinct parts, an external, middle, and internal; the one placed externally is designed to collect and transmit the sonorous vibrations which are modified in passing along the meatus or middle portion to the internal ear. It is within the cavities of this third portion, protected by the hardest bone of the body, the petrous portion of the temporal bone (and on that account called petrous or *stony*), that the nerve destined to the perception of the sense of sound exclusively resides.

659. The concha and the meatus may be compared to an ear-trumpet. The concha contains several prominences or reflectors corresponding to corresponding depressions calculated to concentrate the oscillating rays, and it consists of a thin, tough, and elastic tissue designed to reflect sounds and to increase their strength and intensity by the vibrations to which it is liable. The cartilage of the ear is covered by a very thin skin with no intervening fat which would impair its vibrating and reflecting qualities, and its prominencies are connected together by small muscles, which vary its direction and place it in unison with sounds. In the timid hare, whose only defense consists in flight, the muscles of the ear act in response to sound with the swiftness of lightning, and they vary the direction of the ear towards the quarter from which noise proceeds, so as to meet and distinguish the slightest intimations of danger.

660. The cavity of the tympanum is filled with air,

and this is being continually renewed by the Eustachian tube, which is a half bony and half cartilaginous tube extending from the upper part of the throat to the cavity of the tympanum. Small muscles attached to the three bones move these—tighten or relax the membrane of the drum to which they are attached—and thus institute a due relation between the organ of hearing and the sounds which strike it. It is easy to conceive that the relaxation of the membrane of the tympanum or drum weakens acute sounds in the same manner as slacking the cords deaden the sound of the instrument after which it is named, and conversely if the membrane is tightened.

661. In the same manner as the eye by the contraction or dilatation of the pupil accommodates itself to light, so as to admit a greater or smaller number of rays according to the impression which they produce, so by the relaxation or tension of the membrane of the tympanum, the ear reduces or increases the strength of sounds whose violence would affect its sensibility in a painful manner, or whose impression would be insufficient. The iris and the muscles of the tympanum are, therefore, perfectly analogous in their action, and there is as close a connection between these muscles and the auditory nerve as there is between the iris and the retina.

662. The air which fills the tympanum diffuses itself over the mastoid cells, which not only augment the dimensions of the tympanum, and the force and extent of the vibrations which the air experiences within it, but place the vibrations in contact with the solid vehicle of sound—the petrous portion of the temporal bone. And together with this expansion, the vibrations of the membrane of the tympanum are communicated to a thin

membrane which covers the fenestra rotunda, which on its under surface is in immediate contact with a fluid that fills the cavities of the labyrinth, and in which lie the soft and delicate filaments of the auditory nerve —the nerve of hearing; and the agitation of the fluid within the labyrinth determines the sensations of sound.

663. Any single impulse communicated to the auditory nerve, is sufficient to excite the momentary sensation of sound. But most usually a series of impulses or vibrations is concerned, there being but few sounds which do not partake to a greater or less extent of the character of a *tone.*

664. The *tone* of sound depends upon the rapidity of the oscillations—on the number in a given time. The tone produced by a string or other sonorous body that vibrates quickly, is termed *acute* or *sharp* when compared with one that vibrates more slowly; and the latter, when compared with the former, is said to be *grave.* The gravest sound appreciable is considered to result from about thirty-two vibrations per second, and the most acute has been estimated by Savart to be as high as forty thousand per second.

665. If the vibrations of sound excited in a sonorous body be all performed in equal times, they produce a simple and uniform sensation called a *musical tone.* But if the vibrations are various and irregular, they excite a harsh impression, as if various sounds were heard together, causing a disagreeable noise, denominated *discord.*

666. *Concord* or *harmony* result from the agreement of the vibrations produced by two or more sonorous bodies in such a manner, that some of the vibrations of each strike upon the ear at the same time. If, for example, the vibrations of one sonorous body occupy

double the time of another, the second vibration of the latter will strike upon the ear at the same instant as the former. And this constitutes the concord or harmony of an *octave*. Other tones, which, on being struck together cause discord, if struck in succession produce *melody*.

667. The power of discriminating *tones* is very analogous to that of distinguishing colors. And while some persons are endowed with the former, which is commonly characterized as the "musical ear," to such a high degree as to render them uncomfortable if a discordant sound is produced in their presence ; there are other persons wholly destitute of any such power of discrimination.

CHAPTER XXI.

OUR FUNCTIONS AND FACULTIES.

668. In the various phases of society there are many subjects combining more or less of science, with the most imaginary and speculative fallacies; some to be established, some to disappear, and some to recur again and again, like ill weeds that cannot be extirpated, yet can be cultivated to no result profitable to mankind. All persons hear and talk of such things; but all do not examine them. Some decide upon propositions without exercising their judgments, and these may be said to *decline* an examination of what is offered them. But there is another class of persons who endeavor to improve themselves by an appreciation of the works of their colaborers and of other men; they apply their senses and compare propositions with the accepted laws of nature, which are to them the first tests of all new pretensions. The difference between these two classes, namely, those who endeavor to satisfy their minds on what is offered them by a comparison with accepted truths, and those who accept propositions without examination, is very great; the former only, add to the general stock of knowledge, and advance the cause of science—they are *investigators;* the latter are obstacles to scientific progress, because they do not distinguish between truth and falacy; such per-

sons are only *presumers.* And it is not a little remarkable, how well these latter distinguish themselves by the unreasonable manner in which they challenge and taunt the former for not publicly noticing their presumptions ; and they try to press this circumstance into a tacit acknowledgment of their pretensions.

669. The first and last step of true learning is *humility*, and the scientific person is always advancing in that knowledge which makes him aware of his own wants ; no shadow of perfection ever crosses his mind ; and when he has done all he can, he still has errors to bemoan, and feels more and more the imperfection of his attainments.

670. An individual who asserts anything new, has no right to claim a yes or a no, or think, because none is forthcoming, that he is to be considered as having established his assertion. So much is unknown to the wisest man, that he may often be without an answer, as frequently he is in the region of hypotheses and not of facts. An individual who makes assertions, or draws conclusions regarding any given case, ought to be competent to investigate it. He has no right to throw the onus on others, declaring it their duty to prove him right or wrong. His duty is to demonstrate the truth of what he asserts, or to cease from asserting. The men he calls upon to consider and judge his assertions, have enough to do with themselves, in the examination, correction, or verification of their own views. I am not bound to explain how a table tilts, any more than to indicate how, under the juggler's hands, a cabbage appears to be turned into pudding. The means are unknown to me. Let those who affirm the exception to the general laws of nature work out the experimental proof.

671. When scientific persons are led by circumstances to give an opinion adverse to any popular notion, nothing is more common than the attempt to neutralize the force of such an opinion, by reference to the mistakes which like educated men have made. And their occasional misjudgments and erroneous conclusions are quoted, as if they were less competent than others to give an opinion. Sir Humphrey Davy is said to have expressed the opinion that gaslighting was impracticable on a large scale. Ever since the converse of this opinion has been realized, speculators have used Davy's mistake as a means of tempting moneyed men into companies, or into the adoption of popular fancies; as if an argument were derivable from that in favor of an object to be commended! Why should not men learned far beyond their traducers be expected to err sometimes, since the very knowledge in which they are advanced can only terminate with their lives? There is nothing about them which pretends to perfection; they cannot learn all things, and they may often be ignorant. But the very progress which science makes amongst them as a body, is a continual rebuke and correction of ignorance; that is, of a state which is ignorance in relation to the future, though wisdom and knowledge in relation to the past.

672. If we are to estimate the utility of a science or profession, do not let us hear merely of the errors which have been corrected by others taught in the same careful school; but let us compare that which, as a body, they have produced, with that supplied by their traducers.

673. Where are the established truths of mesmerism, table talkers, clairvoyants, and the host of other delusive and misleading pretenses which are heralded forth

as revolutions? What one result in the numerous divisions of science or its applications can be traced to these "reformers" and pretenders? Where is the investigation completed, so that as in gaslighting all may admit that the principles are established beyond question? Or to light, as used in its application to photography! How wonderful are the results! Light is made to yield impressions upon the metallic plate or the coarse paper, beautiful as those produced upon the living retina; it is made to give a result to be seen now or a year hence—to paint all natural forms and colors; it serves the offices of war, of peace, of art, science, and economy; it is made to take the place of *intelligence*, for a little lamp is set down and left to perform the duty of watching the changes of magnetism, heat, and other forces of nature, and to record the results in pictorial curves, which supply an enduring record of their most transitory actions. And of electricity. In the hands of the careful investigator, what has it not achieved? It descends from the atmosphere, bursts from the metal, dives to the bottom of the sea, surrounds the globe—it talks, it writes, it records in ineffaceable LETTERS OF LIGHT the rewards of science. In all these things, and the catalogue might be extended, what have the presumptuous "new lights" contributed? They profess to deal with subtleties and agencies far more exalted than light or electricity; they lift tables, turn hats, *look at* the internal organs of the human body and describe their appearance, or look into the next room, town, country, and into futurity! Why did they not inform us of the possibility of the photograph? or, if they did not think of it till it was known to scientific men, why did they not then favor us with some instructions for its improvement?

They deal with mechanical forces as well as "spiritual," and they employ both the bodily organs and the mental. Yet, with all these agencies and multiplied powers, they have added no one discovery to the elucidation of a chemical law; they have not added one metal to the sixty-five known to chemists; they have added nothing by their "internal inspections" to the science of physiology; they have not added one planet to the constantly increasing number under the observant eye of the astronomer, and they have not corrected one of the *mistakes* of the philosophers, whose laws they set at defiance. *Not one* new power has been added by them to the means of scientific investigation; *not one* utilitarian principle established.

674. All persons can find in the study of natural phenomena a school for self-instruction and an illimitable field for the exercise of the mind. The bare contemplation of nature's laws inspires a respect for truth, while it invites investigation; and when we arrive at the full appreciation of one of these laws, we find it to be no less precise than comprehensive, that it has been curtailed of all useless complications; there is about it no redundancy, no verbiage, no mystery. And the *laws of nature*, as thus understood, are the first tests to which all new theories or declared "facts" should be subjected, ere they are received as truths. Pre-eminent among the laws of nature stands the law of gravitation, and its undeviating truth stands as enunciated by Newton—that matter attracts matter with a force inversely as the square of the distance. Newton showed that, by this law, the general condition of things on the surface of the earth is governed, and the globe itself, with all upon it, kept together as a whole. Yet more, that the motions of the planets round the sun, and of

the satellites about the planets, all bore testimony to this great truth in their operations under its government. Numerous discoveries of many great naturalists since Newton have added the testimony of worlds many times larger than ours, rolling in the depths of space under the government of Newton's law. Yet this law, under whose benign influence all nature harmoniously exists and moves, is made to yield to *table-turning* and *kindred sophisms.*

675. In these days of "progress," persons profess to overcome gravitation by the "affinity" of their fingers. That, by placing their fingers on a table and then raising their hands, the table straightway gives up allegiance to the law of gravitation, and follows the new light in accordance with the "spirit of the age!" That the table, though heavy, *ascends* and follows after the charmer's fingers, which bear no weight and are not drawn down by the ponderous mass!

676. "The parties who are thus persuaded," observes Farady (*Lectures on Education,* London), "and those who are inclined to think and to hope that they are right, throw up Newton's law at once, and *that* in a case which of all others is fitted to be tested by it, or if the law be erroneous, to test the law. I will not say they oppose the law, though I *have* heard the supposed fact quoted triumphantly against it; but as far as my observation has gone they will not apply it. The law affords the simplest means of testing the fact, and if there be, indeed, anything in the latter new to our knowledge (and who shall say that new matter is not presented to us daily, passing away unrecognized?), it also affords the means of placing *that* before us separately in its simplicity and truth. Then why not consent to apply the knowledge we have to that which is

under development? Shall we educate ourselves in what is known, and then, casting away all we have acquired, turn to our ignorance for aid to guide us among the unknown? If so, instruct a man to write, but employ one who is unacquainted with letters to read that which is written; the end will be just as unsatisfactory, though not so injurious, for the book of nature, which we have to read, is written by the finger of God. Why should not one who can thus lift a table proceed to verify and simplify his fact, and bring it into relation with the law of Newton? Why should he not take the top of his table (it may be a small one), and, placing it in a balance, or on a lever, proceed to ascertain how much weight he can raise by the draft of his fingers upwards; and of this weight, so ascertained, how much is unrepresented by any pull upon the fingers downwards? He will then be able to investigate the further question, whether electricity or any new force of matter is made manifest in his operations; or whether, action and reaction being equal, he has at his command the source of perpetual motion. Such a man, furnished with a nicely constructed carriage on a railway, ought to travel by the mere draught of his own fingers. A far less prize than this would gain him the attention of the whole scientific and commercial world, and he may rest assured, that if he can make the most delicate balance incline or decline by attraction, though it be only with the fourth of an ounce, or even a grain, he will not fail to gain universal respect and most honorable reward."

677. All natural phenomena take place in a regular manner, that is, according to established order or law. And it is the purpose of every science to discover the nature of the laws which govern it. In comparing

phenomena of any kind, for the purpose of arriving at a principle common to them all, it is necessary to feel certain that all the phenomena contemplated are of a similar character; otherwise they can never be classified.

678. It is recorded of Newton, that while contemplating the simplicity and harmony of the plan according to which the universe is governed, as manifested in the relations which his gigantic mind discovered between the distant and comparatively unconnected masses of the solar system, his thoughts glanced towards the organized creation; and reflecting that the wonderful structure and arrangements which it exhibits presents in no less a degree the indications of the order and perfections which can result from Omnipotence alone, he remarked, "I cannot doubt that the structure of animals is governed by principles of similar uniformity." The brilliancy of Newton's genius was shown in the perception that the fall of an apple to the earth, and the motion of the moon around it, were comprehensible under the same law. It is by the exercise of skill like this that the greatest philosophers have been able to achieve their triumphs in the reduction of facts under the dominion of general laws.

679. The apparent dissimilarity of living things which are made the objects of comparison in trying to arrive at the laws which govern their existence, tends to prevent their true relations from being detected. But if we have been diligent students of the *conditions* of vitality as discussed in the preceding pages, we should now have such *clear ideas* of the laws which promote or retard life in all its forms, as to be in a measure capable of reconciling the apparent dissimilarities of living things with the varied circumstances of their development and growth.

680. We have seen that life, in its simplest manifestation, resides in the single cell; and that this cell, when surrounded by congenial elements, is endowed with the function of reproduction by subdivision. It is thus that new cells are produced under the influence of life; and this process of reproduction is continually going on according to the conditions of the surrounding medium in which the cells are generated. The favorable or unfavorable influences determine whether the already existing parts or cells persist, increase, or diminish; whether the qualities necessary to the varied play of the organic functions remain, or whether the machinery of life is arrested by the change or cessation of these qualities.

681. Every cell, every component part of the most complex organic body, has an individual life of its own, limited in duration; yet every such component contributes to the organization and support of the life of the whole being of which it forms part; and the adaptation of life to a complex organism depends upon the due performance of the vital endowments of all the living cells of which the organism is constituted. The different component parts of every organized body are thus, to a certain extent, independent of each other; while, together, they form a nicely balanced whole, the particular parts of which vary only within certain limits.

682. At every instant, and under all circumstances, an organized body offers the sum of its vital operations to the advantage or disadvantage of health, according as it is affected by the vital condition of its constituent parts. So that all irregular actions, disturbances, and pains which follow are just as much in accordance with the laws which govern the existence of a single

cell, as if the cell relied on its own independent action. And all the changes that take place, whether favorable or unfavorable to the continuance of life, are based upon the same fundamental laws.

683. Every organic body is so constituted as to have a time of development and growth ; a period of middle life, in which the functions strive to maintain an unaltered mass ; and an epoch of decrease or decline, which is concluded by natural death. These periodical changes are foreshadowed in the life of the simplest living cell ; it, too, has a period of development, a period of maturity, and a period of decline ; and it is only in accordance with these conditions, that life is embodied.

684. The nature of high development is such, that a most intricate connection is established between the organic functions, and this connection has a constant relation to the necessity of harmonizing the functions with each other, and keeping them in sympathy. And the constant physical and chemical changes which accompany life depend upon the various reciprocities which are produced by the work of the different parts of the body ; the assimilation of what is received, the elimination of what is useless, and the restoration of the organs, by which these operations are effected.

685. In the series of functions observed in the human system, that of DIGESTION, elaborates or separates the nutritive material from alimentary substances ; ABSORPTION provides for the transition of whatever is to be added to the blood ; CIRCULATION sends the blood throughout the system, in order that it may be renovated and applied to the maintenance of the body ; RESPIRATION and CUTANEOUS TRANSPIRATION effect the exchange of gases, and graduate the temperature of the body by the ex-

halation of watery vapor; NUTRITION maintains, increases, or diminishes the mass of the constituents of which the entire organism is composed; and EXCRETION throws off all worn-out and useless matters, the retention of which in the blood would poison the body. Meanwhile, the SENSES receive impressions of the external world; MOTION, with its attendant phenomena, leads to the change in space of particular parts or the entire mass; SPEECH, the peculiar gift of man, which enables him to communicate with his fellows; and, crowning all these, INTELLECTUALITY, the highest function of the nervous system, that function by which we are capable of discerning the beauties of our mechanism and the endowments of our nature! Between all these actions there is the closest bond of union and a coöperation of the component parts of every organ, and every system of organs for fulfilling the purposes of life.

686. We are not so wonderfully made that our existence should be merely maintained. Our organization indicates a nobler and a higher purpose. We need no reasoning to convince us that an organism, so curious and so wonderfully perfect in all its parts as the human, was designed to continue as long as the material composing it will admit of, and that upon us devolves the duty of giving it that continuance.

687. Our duty is known from our nature. What we ought to do for ourselves is as fully understood by a knowledge of our organization and powers, as the uses of any machine is understood by an acquaintance with its construction.

688. The preservation of health is an incumbent duty. We must preserve it in its most perfect state, that in which the powers of the constitution can be best exerted. All the health and strength of which we are

capable, were intended for use ; and any unfitness for the functions of life is a partial death, by a suspense of the compensating powers of the system. The life and activity of every part is merged into an organism so perfect, that all the organs composing it are united together in a bond of mutual dependence, and the complete performance of the *entire series of actions*, is necessary for the healthy maintenance of any one action. All the functions are so completely bound up in each other, that none can be suspended without seriously embarrassing or causing the cessation of all the rest. Hence, if any one organ is diseased all the others come to the rescue ; and some of them assume the functions of the disabled organ until health is restored.

689. The functions of nutrition and elimination are so adapted to each other in the adult system, that the corporeal mass is never suddenly altered, but income and expenditure are very nearly equal So that we have a regular clock-work, which is always correct within a certain limit, and which strives to maintain its ordinary rate in spite of many disturbances.

690. We are all so placed, that there are very few of the objects surrounding us which may not be serviceable or hurtful ; nor is that service to be obtained or injury avoided, otherwise than by our acquaintance with things external, and their relations to our existence. The more exact our knowledge of this kind is, the more we lessen the calamities and add to the comforts of life.

691. That the more the mind is informed, the better it can direct us is evident to all. The mind is, indeed, as much assisted by knowledge as the eye by sight. And whatever the intellectual powers, they are to the individual according to the application of them ; and

in the same manner as that the advantage received from sight is according to the use made of it.

692. The activity of the mind is as much the result of consciousness of external impressions, as the life of the body is dependent upon the appropriation of nutrient materials. But there is this difference between the two : that, while the body continually requires *new* materials and a continued action of external agencies, the mind when once called into activity remains active, and may unfold new relations after the complete cessation of sensations which have served to store it with ideas. The mind thus feeds upon ideas which it has laid up during the activity of functions which may have been destroyed. Not only are the impressions which the mind retains worked up into a never-ending variety of combinations, affording new sources of action to the end of life, but impressions of which the mind, though once conscious, has lost the trace or "forgotten," may recur spontaneously and influence trains of thought at periods long subsequent to their reception.

693. Every impression of which we become conscious is ineffaceably stamped upon the mind, and perpetuated in such manner as to allow of its recurrence to the memory at a future time when it may be roused to action. Examples of this kind are occasionally furnished by persons in delirium :—There is a case on record of a woman, who, during the delirium of fever, continually repeated sentences in languages unknown to those who were around her, which were found to be in Latin, Greek, and Hebrew, chiefly of the Rabbinical dialect. Of these, she stated on recovery, she was totally ignorant. But on tracing her former history, it was ascertained that in early life she had lived in the family of a clergyman, who had been accustomed to

walk up and down his passage, repeating or reading aloud sentences in these languages, which she must have retained in her memory unconsciously to herself.

694. Thus it is that when once the mind is roused into the exercise of its faculties, no subsequent suspension of the power of receiving sensations can again reduce it to a state of inactivity. Sensations and ideas once in active existence are stored up in the mind, which, by the power of memory, "feeds upon the past."

695. There are certain ideas which spring up in the mind during the course of its operations, which are so universally present, that they may be regarded as fundamental axioms of thought. Of such, are the belief in our *present existence*, or the faith which we repose in the evidence of consciousness; this idea being necessarily associated with every form and condition of mental activity. The belief in our *past existence*, and in our *personal identity*; our idea of *time* and *space* as depending upon a belief in the independent existence of the causes of our sensations; the belief in the existence of *an efficient cause* for the changes which we witness, from which we derive our idea of *power*; the belief in the *stability of the order of nature*, or the invariable sequence of similar effects from similar causes; the belief in our own *free-will*, involving the general idea of voluntary power as a result of internal perceptions. These ideas are universally present to thinking minds; they are the chief source of our knowledge, and constitute the sum of our *judgment*. And it should be constantly kept in view, that these fundamental axioms of the intellectual faculties are the expressions of the general truth, that the ideas in question are uniformly excited in all ordinarily constituted minds, by simple *attention* to the changes which take place in natural

phenomena. Hence, we may safely conclude that, notwithstanding the diversities of human character and mental endowments, there are certain *fundamental uniformities* which all alike possess.

696. There is a certain season when our minds may be enlarged, when a vast stock of useful truths may be acquired, when our passions are easily controlled, when right principles may be so fixed in us as to influence every subsequent action of our lives. The season for this is during the youthfulness of intellectual vigor, and if neglected at this period, ignorance and misconduct are, according to the ordinary course of things, the common result. And wrong inclinations become so confirmed, that they defeat all subsequent efforts to correct them.

697. To whatever extent we may be ready to admit the dependence of our mental operations upon the different degrees of organization and functional activity of the nervous system, there can be no question that the *degree of our intelligence* wholly depends upon our sense-perceptions. To our senses we trust directly, and by them we become acquainted with external things and gain the power of increasing and varying facts upon which we entirely rely for all the information we possess or can acquire. It is by our senses that the mind is instructed through every step of life. Intimations flow in upon the mind in proportion to the use made of the senses, and they are stored up in the memory as so many data, and often without our being conscious of them, for the formation of the judgment.

698. The whole theory and practice of education involves the distinct recognition of external influences, as having the most important share in the formation of character. And it is the object of every enlightened

educator to promote the right exercise of that power by which each individual becomes the director of his own conduct—the arbiter of his own destinies.

699. "I consider a human soul," says Addison, "without education, like marble in the quarry, which shows none of its inherent beauties till the skill of the polisher fetches out the colors, makes the surface shine, and discovers every ornament, cloud, spot, and vein that runs through the body of it."

700. It is the trait of our age that, from one end of the country to the other, we are as busy as bees, grinding away at the surface of the rising generation, searching for the ornaments, exhibiting the clouds, and fixing the spots and veins that run through the body of that living marble, from which it is the joy of the artist to clear away superfluous matter and to remove rubbish, in order that the great, the wise, and the good may be disinterred and brought to light. And until we discipline the mind of the multitude, give it the power and the habit of governing itself as well as teaching itself, release it, so to speak, from the iron bondage of utter ignorance of the laws of life and health, and withdraw it from the veil of utter darkness, till we can reach the moral sense and inspire poverty with self-respect, self-knowledge, self-reliance, and the courage to pursue paths open to the lowest as well as the highest; not until we have done this shall we have achieved the true purposes of our nature.

CHAPTER XXII.

THE SUM OF LIFE.

701. The creatures below us are wholly engrossed with the pleasures of sense, because they are capable of no higher existence. But as man was destined for greater and nobler pursuits, he should exercise his capabilities, employ his better nature.

702. When the body is in full health and strength, the mind is so far assisted thereby that it can bear a closer and longer application. Apprehension is readier, imagination livelier, the compass of thought is more capable of enlargement, perceptions can be more quickly examined and more exactly compared, and a truer judgment can be formed. We can in all things have a clearer understanding of what is best for us, of what is most for our interest, and thence determine ourselves more readily to its pursuit, and persist therein with greater resolution and steadiness. It is in this way that the soundness of the body is serviceable to the mind—each needs, each helps the other.

703. The mind, when not restrained by ill health of the body, can, with much greater facility, prevent that discomposure and trouble by which bodily health is always injured, and preserve that quiet and peace by which health is promoted. Hence we should avoid not only that which necessarily brings on disease, but what-

ever contributes to enfeeble and enervate us ; not only what has a direct tendency to shorten life, but likewise whatever lessens our activity, whatever abates our vigor and spirit. We must be intent on such a care of our health as will procure us the *fullest* use of our frame, as will enable us to receive from it the whole of the advantages it is capable of yielding, and so exercising the members of our body, consulting its powers and supplying its wants, that it may be the least burdensome to us, and give the least uneasiness. We should also so guard against the impressions of sense and deceptions of fancy, that our faculties may be in a condition to discern what is most becoming and fit for us.

704. If, during the period of youth, we are careful to maintain the mutual relations of the functions of the body and the faculties of the mind, we shall on the attainment of mature years be fully prepared to enter upon the responsibilities of adult age. And we shall be fortified against all the incidents common to the most varied occupations, without fear of encountering any influences essential to their pursuit.

705. All grades and occupations in life have their casualties, and some occupations are essentially dangerous and unhealthy, while they are, nevertheless, highly commendable. The soldier and sailor proverbially lead dangerous lives. The traveler and the geologist are liable to casualties which make life uncertain. The chemist may inhale noxious gases, or endanger life through the heated atmosphere in which he seeks discoveries fraught with benefits to mankind. The anatomist breathes poison from the dead, that he may learn how to protect the living. The clergyman, in the pestilential masses through which he labors with a blessed

perseverance, imparting hope and consolation to the dying, while he reclaims the living, is liable, in the midst of his heroism to be struck down by the diseases which the vices and recklessness of those he is assisting in his Master's service, have brought down upon themselves. And the patient scholar may wear out in intellectual labors and pass to the grave, having enjoyed the less of this world that the more may be left behind to adorn and dignify it. These, and others might be added, are examples of the highest uses of our powers for the benefit of our race.

706. But the ordinary occupations seldom involve such peril of life, as not to be remedied by the application of precautionary measures. The miner is now protected by the safety lamp of Sir Humphrey Davy, from what was at one time regarded as an inevitable peril of his labors. The causes of ill health are as easily removable from most other dangerous occupations, and it is usually more owing to the negligence of precautionary measures than to anything in the nature of an occupation that diseases are occasioned, which are popularly believed to be inseparable from particular pursuits.

707. Everything else being equal, there is no question but that those occupations which are conducted in the open air and in the clear light of day, are of all others, most healthy. The necessity of a sufficient quantity of pure air and light at all times, and under all circumstances, applies to every period and to every object of life. Wherever air and light are deficient, respiration is necessarily imperfect, and no matter what the occupation, or where the abode; however well chosen and nutritious the food; however minute the attention paid to cleanliness; with whatever care the

clothing be adapted to temperature, or how well the duration of exercise, sleep, and waking be regulated; if our houses or rooms be so placed, that the sun's rays cannot enter them, or the air cannot be renewed with facility, consumption and scrofula will inevitably supervene. And the earlier in childhood these causes are applied, the more certainly will disease be induced. Up to the period of the full development of the system, till the body has ceased to increase in stature, till it has reached maturity, and acquired the stability of adult age, the dangers from deficient air and light are much enhanced. After mature years, if the system has been well fortified by a due observance of the conditions most promotive of health, the powers of resisting the causes of disease are very much increased.

708. Literary occupations are generally supposed to be unfavorable to health, and such is perhaps sometimes the case. But it is only when they are rendered so by the conditions already considered. The studious may be so overfond of books, and become so enraptured with the delights of study as to neglect physical exercise, nutrition, sleep, etc. And the necessities of literary persons may incur the ill effects of the same causes, but such coincidents are wholly foreign to literary pursuits, and they have much less relation to the student than to many other occupations, which, because they are deemed less honorable than "hard study," are rarely alleged as causes of ill health. It may be consolatory to relatives to have this cause assigned, and there may be something soothing in the idea that self-sacrifice has been voluntarily incurred by a laudable ambition, which has been esteemed so creditable in all ages, but especially in youth. And the suffused eye may be made to gleam with melancholy pleasure, in re-

flecting upon the honorable path which the unfortunate victim to the misnomer of an over-crowded school-room, deficient exercise, or improper clothing has incurred. But *study* is too sacred, too useful, to be thus defamed. And so far from studious habits being detrimental to health, the contrary is the case; it may be safely affirmed that literary occupations are *promotive* of health and favorable to long life. And when health seems to suffer from intellectual labors, the evil may be remedied by a proper attention to circumstances, having no necessary relation to the pursuit.

709. In a number of the *London Quarterly Review*, published some fifteen years ago, is a list of twenty names from among the most celebrated female authors, whose united ages amount to 1492 years. The youngest being forty-seven, and the eldest ninety-three, the average age of the whole twenty being seventy-four and a half years. Similar lists have been published of others standing high in literature and science, from which Dr. Madden, author of *Infirmities of Genius*, deduced the following order of longevity and the average duration of life of the most eminent in their respective pursuits, taking twenty of each.

TABLE OF LONGEVITY.

	Years.	Average years.
Natural Philosophers	1494	75
Celebrated Female Authors	1492	74½
Moral Philosophers	1417	70
Sculptors and Painters	1412	70
Authors on Law	1394	69
Medical Authors	1368	68
Authors on Revealed Religion	1350	67
Philologists	1323	66
Musical Composers	1284	64
Novelists and Miscellaneous Authors	1257	62½

	Years.	Average years.
Authors on Natural Religion	1245	62
Dramatists	1244	62
Poets	1144	57

710. It may be safely concluded from this table, that literary pursuits are favorable to longevity, and that study promotes it in the same degree as students tend to become scholars. It must be borne in mind, however, that literary occupations cannot be closely pursued with the same intensity in youth as they can in adult age. If, during the period of youth, any organ be immoderately used, a larger afflux of blood takes place towards it, and its vital activity increases in such a manner as to involve a constant overstraining of a particular function; this condition disturbs the balance of power essential to healthy development. Hence it is easily perceived, that the too exclusive exercise of the intellectual powers may occasion an increased flow of blood to the brain, and lay the foundation of disease there, or result in weak bodily organization. Nothing, indeed, seems to be clearer than that full intellectual development requires that the different corporeal functions should be regularly exercised. It is impossible for the mind to aspire to lofty conceptions, or for the various intellectual faculties to be fully accomplished, unless the body is devoid of suffering. Whatever distracts the mind from its own functions, enfeebles its power, and nothing can do this more effectually or unpropitiously than bodily suffering.

711. Every one must have felt the difficulty of bending the intellectual powers on any important topic, when suffering under slight bodily disorder; how much more is this the case under the continued pressure of disease! It may be readily imagined, however, that

13*

although sickness interferes with the vigorous exercise of the mind, it may yet be the occasion of greater productions than a state of health. Disease necessarily confines the invalid; and this confinement incites intellectual exercise for the purpose of dispelling the ennui which such a condition induces. Thus the *production* may in some cases be greater although the *capabilities* may be less.

712. Cheerfulness is an important aid to good health, under whatever circumstances. It banishes all anxious care and discontent, sooths and composes the passions, and keeps the soul in a perpetual calm. In childhood and youth, the full play of the emotions and of the organs on which their effects are exhibited, is indispensable to health. And evil has always been found to result from any plan of education which has not included juvenile enjoyments. Under the play of the emotions, cheerfulness of temper becomes an inherent part of the constitution; and the action of the different functions goes on with such vigor, as to resist all injurious influences.

713. To expect to develop an active mind, communicate graceful form and motion to the limbs, impart health and vigor to the body and cheerfulness of temper by confinement, belts and splints superadded to the lessons of a posture-master, is about as rational as would be the attempt to improve the beauty and vigor of our forest trees by transplanting them to the greenhouse, and extending their branches along an artificial frame-work.

714. Providence has with a bountiful hand, prepared a variety of pleasures for the various stages of life. And the object of all sound education is to raise the mind to its due perfection by giving it a taste for those

entertainments which afford the highest transport, without the coarseness or remorse that attends vulgar enjoyments. Amusement alone is not sufficient for this, hence, we must needs choose those diversions which will permit the greatest freedom of motion and unrestrained exercise of the limbs. Of all amusements adapted to these purposes DANCING is pre-eminent, especially for girls.

715. It is important to secure to both sexes all the physical advantages which nature has formed them to enjoy ; both should partake of the same rational means for ensuring that vigor of the body and cheerfulness of mind, which will most promote the healthy exercise of the functions, and by which a sound maturity can be produced. But custom and propriety alike sanction out-door amusements for boys, which are altogether inappropriate for girls. Dancing has no objection for either. It is, and has been practiced by all nations as an amusement ; and by some nations as a part of religious worship.

716. But there is no physical exercise that requires a more unrestrained play of the respiratory organs than dancing. And many are the fatal results of the pernicious error of tight dressing for the ball-room. No recreation should be indulged, which infringes upon the laws of organization and the dictates of nature. And dancing, if not regulated by moderation and discretion, tends alike to the destruction of intellectual enjoyment, and to the severance of the holiest social ties.

717. It is a wise providence in the plans of the Creator, that the existence of all organized bodies should be temporary. The ages of man are both numerous and protracted. For a time, the parts of the frame that are concerned in his development, unceasingly deposit

the necessary particles by a process as beautiful and systematic as it is mysterious and sublime; until, ultimately, the growth peculiar to the species and the individual is attained. At this point the preponderance which before existed in the action of the exhalants over the absorbents, ceases; all is equality. Ere long the exhalants fall off in their wonted activity; the fluids decrease in quantity; the solids become more rigid; and all those changes supervene which characterize the decline of life.

718. But death may occur at any period of life; a few only ceasing to live by the effects of age alone. The duration of life varies according to numerous appreciable and inappreciable circumstances: the original constitution of the individual, the habits of life, locality, and various other causes sufficiently indicated in the preceding chapters. And although it may not be in our power to comprehend the various causes of disease that exist in nature, yet, by a due observance of the laws by which our functions are performed, we can cultivate experience and observation to that degree of perfection which will teach us how to act prudently and safely.

719. The qualities of natural phenomena all proceed from the same source, and have certain operations, each peculiar to itself, yet all in harmony with every other; so that we have a complexity of result from the simplest means which is eminently characteristic of natural laws.

720. Latitude, elevation, nature of the soil, degree of cultivation, relative position in regard to mountains, forests, rivers, etc., and general aspect of the neighborhood, all modify the condition of man, and prove his adaptability by such effects as serve to make

him understand his relations to what is around him. These relations are uniformly present and do not require lengthy investigation nor instrumental experiments to discover their existence, but are such as can be judged of and acted upon from our sensations alone. We cannot prevent the dews of heaven, nor the heat of the sun, nor the process of decomposition ; but we can understand the course and order of natural phenomena, we can trace out the laws that govern them and ascertain our relations to them. And if we apply our knowledge of the laws of organization in tracing the causes of ill health, it will enable us to escape all such diseases as spring from ignorance and misconduct.

721. The instincts of the lower animals, which regulate their movements in times of danger, are necessary to their preservation. But man, who has the faculty of reason for his guide, and is therefore responsible to the source of reason, must take the consequences of his free agency. As these consequences are invariable, we must learn to accommodate our conduct to them, and to apply known truths in furtherance of practical good.

722. Disease forms no exception to the Divine arrangement of uniformity in physical phenomena. The beginning and end of human life are only steps in an eternal existence. The principle of death is bound up with the principle of life. Death is the completion of life, but if disease had no other purpose nor end than death, it would be an anomaly in the works of the Creator, as involving an arrangement of vitality without salutary tendencies. Like our own handiwork, which has in itself no provision for repair, we should wear out ; labor and sorrow would be the end of all our

days ; life would be a burden, health beyond hope, and eternity a new creation.

723. The Divine institution of disease requires that there should be more or less uncertainty and irregularity in its action. Diversity and dissimilarity are everywhere manifest, and not less so in disease than in the classes, orders, genera, and species of the animals and vegetables of the naturalist ; individuals of the same species are yet strikingly different. The vital and the physical laws of the universe are unchangeable. Human health and disease are alike subject to those laws.

724. The gratification which the reasoning faculties constantly seek, even if it involve a sacrifice of individual health, discloses truth of universal application. And as man tastes of the delights of intellectual action and gives way to the impulses of his nature to pursue them, he will see in disease a providential mercy to encourage his willing submission to it and to mitigate its severity. As its pains are but temporary, he can find strength to bear them patiently, if not to welcome them in the thought of the enduring good which they are intended to work within him.

725. The uncertainty of the issue in any disease, however slight in the beginning, is evidence of its merciful object.

726. Were it otherwise, were we so constituted as never to be sick but unto death, how appalling and hopeless would be the condition of man, hardened in sin by deliberate postponement of immortal concerns on account of the certainty of time ! But the uncertain duration of even the most fatal diseases, guards and secures their fitness for the common purposes, and prevents them from being any exception to the Divine arrangement. Consumption, the most uniformly fatal

of all diseases, is variable and insidious in its outset, fickle in its termination, and frequently cut short by the unexpected invasion of other maladies.

727. But let us go further and suppose disease to be of one kind only, and always fatal at a particular period. Then the case would be much aggravated. A death-bed repentance would be the universal reliance, while health continued there would be no concern for the future state. Feeling sure of time for the necessary preparation, convenience and necessity would take the place of duty, and the deceptions of weakness and bewilderment, instead of strength and clear perception in the full enjoyment of health and faculties would determine the chances of eternity.

728. On the other hand, men sometimes die without the intervention of disease—are suddenly cut off in the full possession of health. But such deaths are rare and exceptions to the general rule. And how would it be if they were the rule instead of the exceptions? How indescribably dreadful would be the fear of certain, sudden death! The whole of life would consist in the dread of impending danger, pleasure would be unexperienced and unknown, and civilization among the things that are not.

729. Indeed, it is only by the present arrangement of disease that its Divine origin can be appreciated and its beneficence discerned. In any other way it would have no analogy to the diversity everywhere manifest in natural phenomena, nor would it serve the merciful purpose for which it was ordained.

730. Disease was not instituted simply as the road to death, or it would have been uniform and certain in its course. True happiness consists in the influence of religion, to which the *whole of life* should be devoted.

That fullness of communion which actuates the most kindly emotions, induces peace, inspires love, and waits for heaven, is more or less the fruit of disease. It brings out and nourishes all the finest feelings of our nature. When strength is laid low and man is made to see and feel his dependence upon his fellow-man, who has watched the skill of the truly Christian physician, and seen the sympathy of his full heart overflow, lest through too much confidence in human aid the purpose of God may not be fully accomplished ; who has seen this and not felt the benefit of the sick room ? And when disease appears in its sternest aspect—when the surgeon has to take the responsibility of hazarding the most intense suffering with the last bare hope of relief —with what sympathy, what self-denying devotion the wife, the mother, the sister, the friend, aye, the enemy, and the most abandoned of mankind, find their feelings softened ! Each and every human soul confesses that it is in mercy and not in wrath that God has sent disease into the world.

731. All conflict with this conclusion is removed by our Lord's blessed answer to the question, " Master, who did sin, this man or his parents, that he should be born blind ? Neither hath this man nor his parents, but that the works of God should be made manifest in him."

732. The apparent punishment of Azariah with leprosy for profanity, and of Gehazi for covetousness and falsehood, was in reality a correction in mercy. And in like manner were the punishments of the Israelites : " When Thy people Israel be smitten down before the enemy, because they have sinned against Thee and confess Thy name and pray and make supplication, then hear Thou in heaven and forgive the sin of Thy people." The purpose of such judgments was not to punish them

for having entered upon a life of sin, but to turn them from the course of it before it was too late, when it might be said of them as of Ephraim, "He is joined to his idols, let him alone;" and of "Those who have forsaken the Lord and provoked the HOLY ONE of Israel unto anger;" "Why should ye be stricken any more? ye will revolt more and more;" "Behold these are the ungodly who prosper in the world."

733. Disease is nowhere manifest as mere punishment, but as a correction in mercy for the salvation of the soul. Even when it is brought about by our own misconduct, it is consonant with this view and therefore salutary, because we are admonished by it to be ready for death.

734. We are corrected against the imprudence committed for future improvement in *ourselves*, not punished that others may profit by our example, which is the true design of punishment. But, as a general thing, there is no connection between acts performed and disease. Indeed, if this were the case, we should find that the righteous and the wicked could be designated by their relative degree of health, and we should be constantly led to inquire, "Who did sin, this man or his parents?"

735. It is conclusive that no manner of life can be alleged in justification of disease. From infancy to old age, the innocent and the guilty are alike subject to its uncertainties.

736. The inherent aversion of man to live for the future, needs a constant monitor. And this is the beneficent purpose of disease. Is it possible to conceive of anything so well calculated to impress the mind with the necessity of a constant state of preparation for death?

737. There is no condition in which we are in so much danger of forgetting that an eternity awaits us, as when we are in the uniform enjoyment of health. The purpose of disease may indeed be disregarded, unheeded; in which case the visitation will be unprofitable, as it was to the Philistines, but its purpose is not on this account altered.

738. However incomprehensible the ways of Providence in this, as in other things beyond the capacity of man to conceive of, we may nevertheless be assured that when God selects the furnace of affliction by which to test and strengthen the graces of the Holy Spirit, there is in it Infinite wisdom. A thorough conviction of God's love and merciful providence in all His dispensations is the only adequate proof of a submissive spirit. And the benefit of disease is to be obtained, not by endeavoring to discover its secret cause, but by meek submission to the Divine will.

739. In the case of Job, the mercy of God in disease is powerfully manifest. He was by his friends thought to be a hypocrite more than they, and therefore more afflicted. "And do ye not know their tokens that the wicked is reserved to the day of destruction? they shall be brought forth to the day of wrath." Job fully exemplifies the purpose of disease to the righteous in his pre-eminent display of integrity, meekness, and piety. And Lazarus, likewise, was "not sick unto death, but for the glory of God."

740. It is thus that the CREATOR, having designed man for a higher sphere, has not only given him the capacity of knowledge and virtue, but has instituted disease as a sentinel on the threshold of his future existence.

INDEX.

A.

B.

C.

D.

E.

F.

G.

H.

I.

J.

L.

M.

N.

T.

U.

V.

W.

Y.

Z.

JAMES COWLES PRICHARD, M.D.

THE NATURAL HISTORY OF MAN; comprising Inquiries into the Modifying Influences of Physical and Moral Agencies on the different Tribes of the Human Family. 4th edition, revised and enlarged. By EDWIN NORRIS, of the Royal Asiatic Society, London. With 62 plates, colored, engraved on steel, and 100 engravings on wood. 2 vols., royal 8vo., elegantly bound in cloth. London, 1855. $10.

Six Ethnographical Maps. Supplement to the Natural History of Man, and to the Researches into the Physical History of Mankind. Folio, colored, and 1 sheet of letter-press, in cloth boards, 2d edition. London, 1850. $6.

SCHLEIDEN.

THE PLANT; a Biography, in a Series of Fourteen Popular Lectures on Botany. Edited by A. Henfrey. 2d edition, 8vo., with 7 colored plates and 17 wood-cuts. London, 1853. $4.

WATERHOUSE.

A NATURAL HISTORY OF THE MAMMALIA. London, 1846–48. Vols. 1 and 2, Marsupiata and Rodentia. 2 vols., royal 8vo., plain plates, $14 50. Colored plates, $17.

J. QUEKETT.

I.

LECTURES ON HISTOLOGY, delivered at the Royal College of Surgeons of England, Elementary Tissues of Plants and Animals. On the Structure of the Skeletons of Plants and Invertebrate Animals. 2 vols., 8vo., illustrated by 340 wood-cuts. London, 1852–54, $5 75.

Vol. II., separately, $4.

II.

PRACTICAL TREATISE on the USE of the MICROSCOPE. Illustrated with 11 steel plates and 300 wood-engravings. 8vo., 3d edition. London, 1855, $5.

REV. J. BERKELEY.

INTRODUCTION TO CRYPTOGAMIC BOTANY. 8vo., illus'd with 127 wood-cuts drawn by the author. Lond., 1857. $5.

BAILLIERE'S

ETHNOGRAPHICAL LIBRARY, conducted by EDWIN NORRIS, Esq., R.A.S. Vol. 1. Native Races of the Indian Archipelago; Papuans. By GEO. WINDSOR EARL, M.R.A.S., Post 8vo., cloth. With 5 col. plates and 2 wood-cuts. Lond., 1853. $2.

Vol. II. Native Races of the Russian Empire. By R. G. LATHAM, author of *The Varieties of Man*, etc. Post 8vo. With a colored map, copied from one made under the direction of the Russian Government, and other illustrations. Lond., 1855. $2.

The books on the following list are marked at the reduced rates. Our terms are CASH WITH THE ORDER, and POSITIVELY NO FURTHER DISCOUNT WILL BE ALLOWED.

It is particularly requested that the manner in which books are to be forwarded should be specified on the order. If BY MAIL, TEN PER CENT. must be ADDED to the remittance, to cover the cost of postage, which the Post-Office regulations require to be prepaid.

BAILLIERE, Bros., 440 Broadway.

Agassiz (L.) An Essay on Classification. 8vo. London, 1859 . . . . . . $3 25

Bossu. Nouveau Dictionnaire d'Histoire Naturelle et des Phenomenes de la Nature. 3 vols., royal 8vo. Paris, 1857, bound in 1 vol. red calf 7 00

Buckland (T. F.) Curiosities of Natural History. 4th edition. 12mo. London, 1859 . 1 62

——— Geology and Mineralogy considered with reference to Natural Theology. New edition with additions. 2 vols., 8vo. London, 1858 . . . . 5 10

Carpenter. Animal Physiology. New edition. Thoroughly revised and in part rewritten. Post 8vo., illustrated. London, 1859 . . 1 50

Chatin (G. A.) Anatomie comparee des vegetaux, comprenant—1o. Les plantes aquatiques ; 2o. Les plantes aeriennes ; 3o. Les plantes parasites ; 4o. Les plantes terrestres. Liv. 1 to 9.

L'ouvrage se public par livraisons de 3 feuilles environ et 10 planches. Il sera complete dans environ 20 liv. Prix de chaque livraison 1 75

Chenu. Encyclopedie d'Histoire Naturelle, ou Traite complet de cette Science, d'apres les travaux des naturalistes les plus eminents de tous les pays et de toutes les epoques. Crustaces, Mollusques et Zoophytes. 4to. Paris, 1859 . . . . . 1 75

Les 18 vols. de cette Encyclopedie publies anterieurement sont.

Quadrumanes $1 75
Insectes Coleopteres 2 vols. . . . 3 50
Oiseaux 6 vols 10 50
Papillons 2 vols. 3 50
Mammiferes 4 vols. 7 00
Botanique 2 vols. 4 00
Reptiles et Poissons 2 00

Clarke. The Microscope; being a popular description of the most instructive and beautiful Objects for Exhibition. 12mo. London, 1858 81

Dalyell. The Powers of the Creator Displayed in the Creation; or, Observations on Life amidst the various forms of the Humbler Tribes of Animated Nature; with Practical Comments and Illustrations. Vol. 3, 4to., containing numerous plates of living subjects, finely colored. London, 1858 . . 11 25
3 vols., 4to. 54 00

Darwin. On the Origin of Species by Means of Natural Selection; or, the Preservation of favored Races in the Struggle for Life. Post 8vo. London, 1859 . . . . . 3 83

Dresser. Unity in Variety as deduced from the Vegetable Kingdom; being an attempt to develope the Oneness which is discoverable in the Habits, Mode of Growth, and Principle of Construction of all Plants. 8vo. London, 1859 . . . . . . 3 00

Edwards (A. M.) Life Beneath the Waters; or, the Aquarium in America. Illustrated by plates and wood-cuts drawn from life. 12mo. New York 1 50
Sent free by mail on receipt of price.

Edwards (Milne). Lecons sur la physiologie et l'anatomie comparee de l'homme et des animaux. Vols. 1 to 5, part 1. Paris, 1858–59 . . . . 9 50
Will be complete in 8 vols.

Flourens. De la Vie et de l'Intelligence. 2e edition. Paris, 1858 . . . . . . $ 75

Forbes (E.) Outlines of the Natural History of the European Seas. 12mo. London, 1859 1 35

Geoffroy Saint-Hilaire (Isidore). Histoire naturelle generale des regnes organiques, principalement etudiee chez l'homme et les animaux. Vol. 3, part 1, 8vo. Paris, 1859 . . . . 1 00

Vols. 1. et 2, 8vo., 1854–59. . . . . 4 00

Godron. De l'Espece et des Races dans les etres organises et specialement de l'Unite de la race humaine. 2 vols. Paris, 1859 . 2 70

Gosse. A History of the British Sea-Anemones and Madrepores. With colored figures of each species. 8vo. London, 1860. . . . 5 67

——— Evenings at the Microscope; or, Researches among the minuter Organs and Forms of Animal Life. 12mo. London, 1859 2 16

Grant. Outlines of Comparative Anatomy. 8vo. Illustrated with 148 wood-cuts. Boards. London, 1833–41 . . . . . 7 00

Hogg (J.) The Microscope. Its History, Construction, and Application. 4th edition, 12mo. London, 1859, ... 1 50

Hogg (R.) The Vegetable Kingdom and its Products. Post 8vo. London, 1858 . . 3 00

Johnstone and Croall. The Nature-printed British Sea-weeds; a History, accompanied by Figures and Dissections, of the Algæ of the British Isles. Vols. 1 and 2, royal 8vo. London, 1859 . . . . 21 50

Jones (Rymer.) Aquarian Naturalist. Illustrated, 8vo. London, 1858 . . . . . 5 00

Jones (J.) The Naturalist in Bermuda. 12mo. London, 1859 . . . . 2 00

Kingsley. Glaucus; or, the Wonders of the Shore. 4th edition, 16mo., illustrated. London, 1859 . . . . 2 00

Man and his Dwelling-place ; an Essay towards the Interpretation of Nature. Post 8vo. London, 1859 . . . . $2 43

Michelet. L'Insecte. 3e edition. Paris, 1859 90

Micrographic Dictionary by Griffith and Henfrey. 2nd edition, 8vo. London, 1860 12 25

Patterson (R.) Series of Ten Zoological Diagrams. Sheet A. Mammalia I., 5 figures. Sheet B. Mammalia II., 6 figures. Sheet C. Aves, 5 figures. Sheet D. Reptilia, 4 figures. Sheet E. Pisces, 10 figures. Sheet F. Mollusca, 23 figures. Sheet G. Arachnida, and Insecta, 12 figures. Sheet H. Insecta, 16 figures. Sheet I. Crustacea, Cirripeda and Annellata, 13 figures. Sheet K. Radiata, 22 figures. Size of each diagram, 40½ by 29 inches. The set of ten, fully colored. On rollers 27 00

Page. Handbook of Geological Terms and Geology. Post 8vo. Edinburgh, 1859 . 1 62

Rainey. On the Mode of Formation of Shells of Animals, of Bone, and of several other Structures. Post 8vo. London, 1858 . 1 25

Recreative Science. A Monthly Record and Remembrancer of Intellectual observation (principally on Natural History). Square 8vo., illustrated. Per annum . 2 50

Richardson. Fauna Boreali Americana ; or, Zoology of North America, containing Descriptions of the subjects collected in the late Northern Expedition under Sir J. Franklin. 4 vols., 4to. London, 1829–36, half-bound morocco, gilt top. Very fine copy . 60 00

Samuelson and Hicks' Humble Creatures. The Earth Worm and the Fly ; in eight letters. Microscopic Illustrations. 8vo. London, 1858 95

West, Tuffen. Half hours with the Microscope as a means of Amusement and Instruction. 12mo.., illustrated. London, 1859 . 68

www.ingramcontent.com/pod-product-compliance
Lightning Source LLC
LaVergne TN
LVHW020214110826
845151LV00003B/707

* 9 7 8 1 4 2 5 5 3 4 3 7 0 *